Jules Verne

Around the world in 80 days

Text adaptation and activities by **Eleanor Donaldson**

Illustrated by **Paolo D'Altan**

Editor: Michela Bruzzo
Design and Art Direction: Nadia Maestri
Computer graphics: Simona Corniola
Picture research: Alice Graziotin

© 2011 Black Cat

First edition: January 2011

DEALINK, DEAFLIX are trademarks licensed by
De Agostini SpA

Picture credits
Cideb Archive; De Agostini Picture Library: 49; Getty Images:
51; Graham Chadwick/Getty Images: 52; Swim Ink 2,
LLC/CORBIS: 88.

We would be happy to receive your comments and
suggestions, and give you any other information concerning
our material.
info@blackcat-cideb.com
blackcat-cideb.com

The Publisher is certified by

 CISQCERT

in compliance with the UNI EN ISO 9001:2008
standards for the activities of «Design and
production of educational materials»
(certificate no. 02.565)

Printed in China by FoShan NanHai XingFa Printing Co.,LTD

Contents

| n. track |
| end |

These symbols indicate the beginning and end of the passages
linked to the listening activities.

About the author

Jules Verne was born in 1828 in the town of Nantes, in France.
When he was a boy, he ran away from home and tried to get on a ship to the Caribbean. The men on the ship found him and sent him back home.
In 1847, Jules's father sent him to Paris to study law. He did not like the subject very much and his father was angry when he left law school and started writing plays instead. His plays were not very successful at the beginning, and he had to find another way to earn money because he was in love with Honorine, a widow [1] with two young children. He became a stockbroker [2] and married Honorine a

---

1. **widow** : this woman's husband is dead.
2. **stockbroker** : a person who helps people invest their money in other companies.

year later, in 1857. They had a son called Michel. During this time Verne continued writing, and in 1852 he wrote a book about how a man could travel across Africa in a hot-air balloon. [3] One publisher suggested that he wrote an adventure story, using the same ideas. He did this, and in 1863, he wrote *Five Weeks in a Balloon*. People liked this new mixture of fact and fiction, and the book was an immediate success.

With the help of his friend and publisher, Pierre-Jules Hetzel, he wrote many books, sometimes two a year. Some of the most famous of these are: *A Journey to the Centre of the Earth* (1864) , *From the Earth to the Moon* (1865) and *Twenty Thousand Leagues under the Sea* (1869). In these stories his heroes are clever men who are able to find solutions to problems and escape from dangerous situations.

This is also the case in *Around the World in Eighty Days* (1873). This was not only Jules Verne's most popular story but he also saw it performed several times as a play during his own lifetime. Many of Jules Verne's stories became classic films, for example, *Twenty Thousand Leagues Under the Sea* (1954).

Jules Verne was popular in his time because of people's interest in science. Today, people are interested to see how many of his imaginary inventions became reality. For example, in *From the Earth to the Moon*, the story is very similar to the real events of man's first journey to the moon in the *Apollo*.

Jules Verne did travel later on in his life, and in 1884 he did a tour of the Mediterranean. He died in 1905 in Amiens, France. Many people think of him as the 'father' of science fiction.

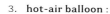 3. **hot-air balloon** :

1 Writing

Write a sentence about Jules Verne for each date on the timeline.

Example: *Jules Verne was born in 1828.*

1828   1847   1852   1857   1873   1905

2 Comprehension check

For each question choose the correct answer — A, B, C or D.

1   Jules Verne's father was very angry when Jules
   A ☐ tried to get on a ship to the Caribbean.
   B ☐ decided to go to Paris.
   C ☐ did not like studying law.
   D ☐ left law school and started writing plays.

2   At the beginning of his career
   A ☐ he changed his mind and became a stockbroker.
   B ☐ he had to marry Honorine.
   C ☐ his plays were not very successful.
   D ☐ he had a lot of money.

3   Jules Verne's adventure books were successful because
   A ☐ people liked this mixture of fact and fiction.
   B ☐ they portrayed the life of heroes.
   C ☐ they were all based on true stories.
   D ☐ his friend and publisher Pierre-Jules Hetzel helped him.

4   Jules Verne was popular in his time because
   A ☐ many of his stories became classic films.
   B ☐ many of his stories became famous plays.
   C ☐ one of his stories was similar to the real events of man's first journey to the moon.
   D ☐ people then were very interested in science.

The Characters

From left: **Mrs Aouda, Mr Fogg, Passepartout**

Before you read

1 **Test your Geography!**

Can you match the places in the photos (A-F) that Phileas Fogg goes to on his journey around the world?

1	Egypt	3	Mumbai (Bombay)	5	Hong Kong
2	Yokohama	4	San Francisco	6	New York

A

B

C

D

E

F

track 02

2 **Listening**

You will hear a description of the places in exercise one. Write the number next to the picture.

3 **Speaking**

Discuss these questions.

1   Have you, or someone you know, been to one of these places/countries?

2   Which places in the world would you like to visit? Make a list. You will look at this question again in Chapter Two.

CHAPTER **ONE**

When Phileas Fogg meets Passepartout

Let me begin by introducing a mysterious English gentleman called Phileas Fogg. Most people don't know very much about him, but because he does the same thing every day, some people think they know everything about him. He is very handsome and he is a true gentleman. He is certainly rich, but no one knows how he made his money.

Has he ever been to another country? He can name a lot of countries on a world map and he knows the most incredible things about them. He probably travelled at one time, but some people insist that he has not left London for many years. Maybe he only travels in his head.

He is a very private man and he does not have many friends. The only time he speaks to other people is at the Reform Club, [1] where he goes to read newspapers and play cards. He does not play to win. He plays for the enjoyment of the game. He often wins, but he does not keep the money. He gives it to charity. He likes to see his games as a challenge; a challenge that does not require any physical effort.

He has lunch at the Reform Club every day, in the same room, at the same table. He goes home at midnight. He lives in his house in Savile Row, a good address in central London. No one ever goes there, except his manservant, who must always be on time and be completely loyal [2] to Phileas Fogg. In fact, this very morning, his manservant lost his job because the water he brought Phileas Fogg was too hot to shave with. And this is where our story begins.

Phileas Fogg was sitting in his armchair waiting for his new manservant at some time between eleven and half past eleven. (At exactly half past eleven Mr Fogg goes to the Reform Club.) He looked up at the hands of the large clock by the wall that counted every second with a loud tick.

There was a knock at the door and a young man of about thirty came in.

'You say that you are French, but your name is John?' asked Phileas Fogg, looking at him carefully.

'Jean, sir, not John,' said the young man. 'Jean Passepartout.

I am an honest man, sir, and I must tell you that I haven't been a manservant all my life. I was a physical education teacher

---

1. **Reform Club** : a political club in London with services for members. It began around 1832 to give members of the Liberal Party a place to meet and discuss their ideas.

2. **loyal** : always supporting someone or something.

and a music teacher; then I became a singer. I once rode a horse in a circus, and for a time I worked for the fire brigade in Paris. I found out that a certain Mr Fogg was looking for a manservant. "He is a very clever, careful man," they told me. "You won't find a quieter man in all of England. He does the same thing every day." And so I came here to ask about the job, in the hope of finally being able to live a quiet life.'

'Yes, someone at the Reform Club told you this I believe — probably the same person who told me about you. Do you understand what type of person I'm looking for?'

'Yes, sir. I do, and I think I'm perfect for the job.'

'Well then, what time is it now?'

'Eleven twenty-two, Mr Fogg,' Passepartout replied, taking his pocket-watch ³ out of a small side pocket.

'Exactly four minutes late,' noted Phileas Fogg, looking at his own watch. 'So, let's say you started working for me as from — eleven twenty-six.'

Phileas Fogg stood up from his armchair, picked up his hat, and went out of the door without saying another word. From this brief introduction, Passepartout was able to make note of his employer. He was about forty years old, an elegant man with an attractive, gentle face. He was tall, with blond hair and a moustache. He was the sort of person who remained incredibly calm, even under pressure. He had gentle eyes that fixed you with a firm stare. ⁴ He never seemed upset ⁵ or worried. He was

3. **pocket-watch** :

4. **firm stare** : to look at someone without taking your eyes away from them.

5. **upset** : unhappy.

a typical Englishman. It was always difficult to guess an Englishman's true feelings.

And our Frenchman? Passepartout had an attractive face and he was incredibly strong. He had blue eyes, and untidy, curly brown hair. He was a sweet person who understood the meaning of true friendship and loyalty.

It was just after half past eleven and Passepartout, who was now alone in his new home, decided to look around. In Phileas Fogg's room his clothes were divided into seasons; each jacket had a number. In the corner, there was a safe [6] for keeping money, watches and other items. After looking in all the different rooms, he finally came to his own bedroom. Above the fireplace there was an electric clock; it was the same electric clock that Phileas Fogg had in his room. The two clocks ticked at the exact same second. Below the clock there was a piece of paper listing the details of Mr Fogg's day.

> 8.20: Tea and toast
> 9.37: Water for shaving (body temperature)
> 11.25: Brush Mr Fogg's jacket
> ...

The list told Passepartout everything he needed to do from morning until midnight when Mr Fogg went to bed.

'Not bad at all,' thought Passepartout. 'A man who is as regular as clockwork! [7] This is just what I was looking for.'

---

6. **safe** : a strong metal cupboard with locks in which you keep valuable things.
7. **as regular as clockwork** : someone who always does everything on time and in the right order.

The text and **beyond**

1 Comprehension check

Read these sentences about Chapter One. Decide if they are correct or incorrect. If the sentence is correct, mark A. If it is not correct, mark B.

		A	B
1	Phileas Fogg made his money travelling around the world.	☐	☐
2	Phileas Fogg gave the money he won to charity.	☐	☐
3	Every day Phileas Fogg went home to a house in Saville Row.	☐	☐
4	It was important to Fogg that his manservant was on time.	☐	☐
5	Passepartout was a manservant for many years in France.	☐	☐
6	Passeparout found it difficult to guess an Englishman's feelings.	☐	☐
7	The clock in Passepartout's room was slow.	☐	☐
8	Passepartout thought his master was adventurous.	☐	☐

2 How quickly can you find the differences?

There are eight differences between the original text and the text below. Can you find them all? Write them below under the text.

In Phileas Fogg's room his clothes were divided into different colours; each jacket had a number. In the wardrobe, there was a bag for keeping money, watches and other items. After looking in all the different rooms, he finally came to his own bedroom. Above the chest of drawers there was an electric clock; it was a red clock like the one that Phileas Fogg had in his living room. The two clocks played music at the exact same time. Below the clock there was a book listing the details of Passepartout's day.

....
....
....

3 Jobs

A Passepartout has done a lot of jobs. Unscramble the words to find jobs he has done. There are also some jobs connected to the author, Jules Verne. Which ones are they?

RAACBOT ....

MEFNARI ....

NSAMTNVERA ....

IMCISUAN ....

NISREG ....

ROSKCTOBKRE ....

ECARTHE ....

RIERWT ....

B Find the jobs in the list above in the word square below. Look carefully! You can read some words backwards. How quickly did you find them all?

N	E	N	T	P	U	M	T	Q	P	M	X
U	T	Y	A	I	O	A	A	O	P	A	T
R	W	R	S	M	B	F	L	O	R	N	X
B	A	F	A	O	E	I	Z	U	E	S	F
K	A	B	R	I	C	R	I	U	H	E	W
E	Y	C	A	E	N	S	I	X	C	R	R
G	A	Y	M	N	B	D	O	F	A	V	I
Q	J	A	F	W	K	Q	R	U	E	A	T
D	N	S	I	N	G	E	R	I	T	N	E
O	W	V	S	N	S	G	R	V	V	T	R
N	A	I	C	I	S	U	M	Q	U	E	K
T	V	W	Q	M	S	I	C	A	U	H	R
X	S	K	H	E	S	M	E	M	H	K	L
D	S	T	O	C	K	B	R	O	K	E	R
U	W	X	C	S	V	F	T	A	A	K	D

4 Adjectives

Match these adjectives with their opposite. Which adjectives can you find in Chapter One? Who or what do they describe?

1	☐ weak	A	tall
2	☐ rich	B	tidy
3	☐ handsome	C	strong
4	☐ stupid	D	ugly
5	☐ untidy	E	dishonest
6	☐ curly	F	clever
7	☐ short	G	poor
8	☐ honest	H	straight

T: GRADE 4

5 Speaking: hobbies/sports

Phileas Fogg played cards with his companions at the Reform Club. He also had an interest in travel. We can say that these are his 'hobbies', the things he does for enjoyment, not for work. Which of the jobs in exercise 3 could be a hobby? Answer these questions about hobbies.

1 Do you have any hobbies?

2 Where do you do these hobbies?

3 Would you like to try a new hobby? If yes, which one?

PET 6 Writing

You find a note on the table from Jane, who lives in your flat. She is at work so she asks you to do some tasks for her today. Leave a note for Jane. In your note you should:

• say what tasks you did

• apologise because you couldn't do all the tasks

• say why you couldn't do the other tasks.

(35-45 words)

7 Routine

Phileas Fogg does the same thing every day. Imagine one day you wake up and you have to do the same thing every day for one year. You have all the money you need. Use you imagination to write a list of things you would like to do every day for a year.

PET **8 Listening**

track 04

For each question there are three pictures. Listen to the recording. Choose the correct picture and put a tick (✓) in the box below it.

1   What is Joe going to do tonight?

A ☐   B ☐   C ☐

2   What time does Carla say she will arrive?

A ☐   B ☐   C ☐

3   What does the woman decide to buy?

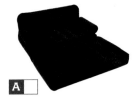  

A ☐   B ☐   C ☐

4   What job have both men done?

A ☐   B ☐   C ☐

17

Before you read

1 Vocabulary

In Chapter Two you will read some words connected to the story of a crime. Use the crossword clues (across/down) to help you find the words for the puzzle. Check their meaning in a dictionary.

robbery   thief   detective   evidence   Scotland Yard
reward   investigation   identity   arrest   steal

Across

1   An .... gives more details about a crime.
3   A ..... takes things that don't belong to him/her.
7   If you find a criminal, sometimes you get a...
9   The crime of stealing money or objects.
10   A .... tries to find criminals.

Down

2   Office for finding criminals in London.
4   Facts to show how a crime happened.
5   Name, personal details, etc.
6   To take someone else's belongings.
8   The police take someone away.

2 Titles

Read the title of Chapter Two. Tick (✓) the sentence which is closest to what you think will happen next.

1   ☐   Phileas Fogg will risk all his money to do something impossible.
2   ☐   Phileas Fogg may lose or win money as a result of something he says he will do.
3   ☐   Phileas Fogg will play a game of cards for a lot of money.

2 October, 8,45 p.m.

CHAPTER **TWO**

When Phileas Fogg makes a bet

Every day, Phileas Fogg left his house at half past eleven. He put his right foot in front of his left foot 575 times — he knew the exact length of every step — and he put his left foot in front of his right foot 576 times before arriving at the steps of the Reform Club.

He usually waited a little before having lunch at thirteen minutes to one. Then he went to the lounge room where he spent the afternoon reading the newspapers. At five o'clock he had afternoon tea [1] and at twenty to six it was time to go to the

---

1. **afternoon tea** : tea with some food, traditionally eaten about four or five o'clock in the afternoon.

Games Room to play cards with other wealthy and respected [2] members of the club, like Sir Ralph Gautier and Andrew Stuart.

On this particular day Andrew Stuart started to read a story to them from the evening newspaper about a robbery at the Bank of England. [3]

The robbery took place on 29 September. The thief stole fifty-five thousand pounds while the head cashier was busy writing a receipt for just a few pence. England's best detectives were looking for the thief after hearing that the Bank of England was offering a reward of two thousand pounds to the person who was able to catch the thief. From the first investigations into the robbery they knew only one thing for certain: he was an elegant, well-spoken [4] gentleman.

While the other members of the club sat at the table, ready to play their game of cards, Andrew Stuart continued to talk about the robbery.

'Where do you think the thief is hiding? He could be anywhere. The world is so big!'

'It isn't so big any more,' replied Phileas Fogg.

'What do you mean?,' said Andrew Stuart with a laugh. 'The earth doesn't get any smaller!'

'Ah! But the earth is smaller,' said Sir Ralph Gautier. 'If you think that we can now go around it ten times quicker than we could one hundred years ago. Did you know that today a man can travel around the world in only three months?'

'Eighty days to be exact,' Phileas Fogg corrected him.

2. **wealthy and respected** : people liked them because of their high position in society, or at the club.
3. **Bank of England** : the central bank of the United Kingdom.
4. **well-spoken** : he spoke good English, like a well-educated gentleman.

'Eighty days?' asked a surprised man at the table.

'Well, maybe that's true, but only if you don't consider bad weather, storms, shipwrecks,[5] and other things,' said another.

'In eighty days, considering all possible events,' continued Phileas Fogg.

'Ah! You think so, do you, Mr Fogg?' laughed Sir Ralph. 'Well, I'll bet four thousand pounds that a journey like that is impossible in such a short time.'

'I repeat that it is possible to do the journey in that time,' said Phileas Fogg, his eyes fixed on Sir Ralph's smile.

'Well, if you are so certain, then do it yourself!'

'I will,' replied Phileas Fogg.

'When?'

'Immediately. And I'll bet not four, but twenty thousand pounds that I can go around the world in eighty days; I will return here in 1,920 hours, or, if you prefer, 115,200 minutes. Do you agree to the bet?'

They all looked at one another. They could not decide if he was serious. 'We agree,' they said.

'Good. I'll take the train for Dover at a quarter to nine this evening. The bet starts as from...' Phileas Fogg took a small notebook and pencil from his pocket and made a note:

2 October, 8.45 p.m.

'And I will return here to the Reform Club at eight forty-five on Saturday 21 December. If I am not here by that time, this cheque for twenty thousand pounds is yours, gentlemen.'

---

5. **shipwrecks** : these happen when a bad storm or another unexpected event destroys a ship at sea.

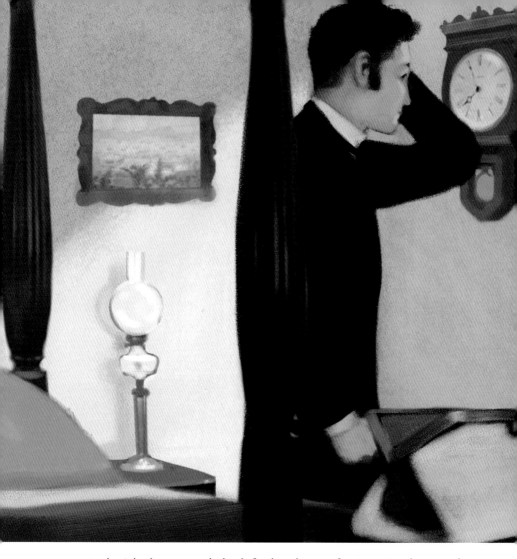

And with these words he left the cheque for twenty thousand pounds on the table, picked up his hat and went out of the door.

At ten to eight his manservant was surprised to see his new employer [6] come through the door. A few moments later Phileas Fogg called to him from his room.

'Passepartout! I need you to prepare our bags. We're leaving in ten minutes.'

6. **employer** : the person you work for.

'Leave home now, Mr Fogg?' Surely his master was not serious.

'Yes,' his master replied. 'We are going around the world.'

'Around the ...?'

'In eighty days,' replied Fogg, 'We're taking a train to Dover. From there a boat leaves to Calais at eleven o'clock tonight. So we haven't a second to lose.'

'Well really!' thought Passepartout, shaking his head. Just when he finally thought he had the perfect job. He wanted to

work for Phileas Fogg because he was a gentleman who lived a quiet life, who always did the same things. And now? How was *this* going to be a quiet life?'

'Pack a small bag with just my night things in it, please, and pack one for yourself. We can buy everything else when we need it,' added Phileas Fogg, and with these orders he left the room.

Passepartout still felt a little confused but he followed his master's orders. He quickly packed their bags and at eight o'clock they were ready to leave the house. Phileas Fogg took a red book showing the arrivals and departures of trains in different countries and a map showing the route of the ships from each port. Then he opened Passepartout's bag, took twenty thousand pounds out of the safe, put the money in Passepartout's bag and closed it tightly. Forty-five minutes later they were on the train to Dover.

Passepartout thought nervously about the notes in his bag. He hoped he was not going to lose the bag.

Phileas Fogg's journey was not a secret for long. The newspapers talked about it with interest and showed pictures of possible routes. Soon everybody in London was talking about Phileas Fogg's departure and his plan to go around the world in eighty days. Some people thought he was mad, others said he was a genius. People were making bets on the possibility of Phileas Fogg completing the journey at all. But a few days later, the front pages had another story. A certain Inspector Fix, a detective for Scotland Yard, said he knew the identity of the thief. All the evidence pointed in one direction: to a well-known and respectable member of the Reform Club — Mr Phileas Fogg.

The text and **beyond**

1 **Do you want to win £20,000?**

How to play: answer question one without looking at the text. If your answer is correct, you win the money. Continue to the next question. If the answer is incorrect, start again. (Check the text for correct answers.) You can ask another student for help with <u>one</u> question, 1-6!

> **ASK A STUDENT ☎ x 1**

1 **£1,000**   2 **£2,000**   3 **£5,000**

4 **£10,000**   5 **£15,000**   6 **£20,000**

1 Who makes a bet of 20,000?

   A ☐ Phileas Fogg   B ☐ Sir Ralphe Gautier
   C ☐ Passepartout   D ☐ Andrew Stuart

2 They read about a robbery. Where does it take place?

   A ☐ at a club   B ☐ in a bank
   C ☐ Fogg's house   D ☐ on the train

3 How much is the reward offered by the bank?

   A ☐ £2,000   B ☐ £5,000
   C ☐ £10,000   D ☐ £20,000

4 On which date will Fogg return to the Reform Club?

   A ☐ 1st October   B ☐ 20th October
   C ☐ 21st December   D ☐ 12th October

5 When does the boat leave for Calais that night?

   A ☐ 10.00   B ☐ 11.00
   C ☐ 09.00   D ☐ 12.00

6 How many steps does Fogg take to reach the Reform Club?

   A ☐ 575   B ☐ 576
   C ☐ 288   D ☐ 1151

2 Summary

Read the newspaper headlines. Number the headlines in the same order you first read about the subject in the text.

1 ☐ **Phileas Fogg: the Gentleman Thief?**

2 ☐ **See the World in Three Months: fact or fiction?**

3 ☐ **Careless Cashier Costs Bank £55,000**

4 ☐ **£20,000 Bet on Impossible Journey.**

2 October, 8.45 pm

Look at the way we say and read dates:

- 2 October: the second *of* October (note: we use *first, second ...* not *one, two...*)
- 8.45: we can say this in two ways — *eight-forty-five* or *a quarter-to-nine.*

3 Dates and times

Answer these questions

1 When is your birthday? Do you know what time you were born?

2 Do you know the birthdays of any famous celebrities?

3 When do you get up at the weekend?

4 What time do you you usually go to school/work?

5 Work with a partner and ask each other questions about your daily routine.
   Ex.: *What time do you usually have breakfast?*
   I usually have breakfast at 7:30. And you?

6 Make a list of well-known events in small groups and then ask each other questions about the dates.

4 Around the World number quiz

Look at the sentences. Write a number in the box for each sentence. When you finish, read out your answers. Write some more questions like these and test someone else.

11,000   8   1.35 billion   6695   7   8848

1   Mount Everest is .... metres high.

2   The Nile is .... kilometres long.

3   The Grand Canyon is sometimes called the ....th wonder of the world

4   The population of China in 2010 was ....

5   The world has .... continents.

6   The Pacific Ocean is .... metres at its deepest point.

5 Game

On a world map, can you find a country beginning with each letter of the alphabet? Which letter can't you find?

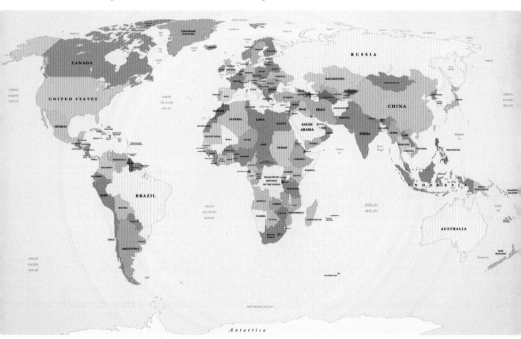

6 **Your trip around the world**

Connect to the Internet and go to www.blackcat-cideb.com. Click on the title of the book and on the Internet project link.

Preparing for your trip

You are going to go on a journey around the world. To prepare for your journey, you need to do some research.

Make a list of the countries you
want to visit

....
....
....
....
....

Use the Internet to find out some information about them:

1   Which cities are there?

2   What places can you visit?

3   Are there any activities you want to do, e.g. elephant rides?

Add to the list things you need to do/buy before you leave.
Number them in the order you are going to do them.

☐ change money at bank   ☐ buy travel guides

☐ find out about visas   ☐ book hotel rooms

☐ get aeroplane tickets   ☐ get passport

✏ ....

Think of three things you would like to take on holiday and why.

Before you read

1 Means of transport

With another person, in five minutes make a list of as many types of transport as you can think of.

2 Reading pictures

Look at the pictures. How many of these types of transport can you find? Use a dictionary if necessary.

> rickshaw   car   bicycle   van   cart   taxi
> lorry   motorbike   train   elephant

3 Prediction

In the next two chapters, Phileas Fogg goes from Bombay (now called Mumbai) to Calcutta. Which types of transport will be used in the text?

CHAPTER **THREE**

When gentlemen
are thieves

Inspector Fix was one of the detectives investigating the robbery at the Bank of England. In his years as a detective he knew only one thing for certain: all the biggest criminals looked like respectable gentlemen. The money, the quick departure. It all made sense. Phileas Fogg was a respectable gentleman, and he, Inspector Fix, wanted to get the reward for catching him.

He soon discovered that Phileas Fogg was on the *Mongolia*, a ship that sailed from Brindisi, in Italy, to Bombay, in India.

Our detective decided to look carefully at all the people getting on and off the *Mongolia*. On Wednesday, 9 October, Inspector Fix saw Phileas Fogg and his manservant as they arrived in the Suez Canal. [1]

---

1. **Suez Canal** : a man-made river, built in Suez, Egypt. It was used by ships to go from Europe to Asia, via Africa.

'So, there's our thief!' he whispered. [2] 'All I need to do now is to tell Scotland Yard and wait for a warrant for his arrest, [3] and then the reward is mine.'

Fix decided to speak to Fogg's manservant.

'Egypt is a beautiful country,' began the inspector.

'Yes, that's true, but we are travelling so quickly,' replied Passepartout.

'Why are you travelling so quickly? Surely you can't see Egypt in only a few days.'

'My master wants to travel around the world in eighty days...' he said, looking at the detective's confused face. 'I know, it's complete madness.'

'Well, your master is... an unusual man, but I imagine he must be very rich to try to do a journey like that in such a short time.'

'To tell you the truth, he has the money he needs. But... I really must leave. We have a boat to catch. Good day, Mr...?'

'Fix, my name is Ins..., Mr Fix. And I believe that we are possibly going the same way. Are you also going to Bombay?'

'Yes, we are. Sorry, not to introduce myself. My name's Jean Passepartout. I'm sure we'll see each other again.' Passepartout touched his hat and waved goodbye.

His conversation with Passepartout made Inspector Fix feel even more certain that Phileas Fogg was the thief. 'I must stop him,' he thought. But how? Without the warrant for his arrest it was impossible, and he could escape again.

2. **whispered** : spoke in a quiet voice.
3. **a warrant for his arrest** : a legal document that gives the detective the permission to arrest someone.

Phileas Fogg, on the other hand, was carefully planning his journey. He kept detailed notes of the date, the length of each part of the journey, the time and the places they stopped in.

On 10 October, the ship left Suez for the next stop — Bombay. The sea was rough, [4] but Phileas Fogg was not worried and soon

4. **The sea was rough** : it was moving a lot.

found people on the ship to play cards with. On 20 October they
arrived in Bombay.

Phileas Fogg seemed to have no interest in exploring the city.
Instead he asked Passepartout to buy some new clothes for the
long train journey. Passepartout was happy to accept. For
Passepartout Bombay was a new and exciting place to be. There
were so many things to buy and there were clothes of every colour.

In the end he bought two sun hats, some new shirts and a new pair of sandals for himself. He was going to return to the station when he saw some girls wearing long red dresses and gold jewellery; they were dancing in the street and people were following behind them. Passepartout continued to follow them until he saw a famous pagoda not far from the road.

'Well, I still have an hour. After all, why should I go around the world and not see anything?' he thought.

Once inside, Passepartout stared in amazement [5] at the strange statues. In fact he was so interested in the detail on the walls that he completely forgot one small detail at the door. One moment Passepartout was looking at the ceiling, the next moment he was staring at the floor. A priest took his arms and another priest took Passepartout's sandals and hit him with them! Passepartout pushed the priests away. The priests fell to the ground leaving the angry people in the pagoda to run after Passepartout, past the sign at the door that read: No shoes inside the pagoda.

Passepartout did not stop running until he arrived at the station with the bags, but he was without his new shoes. He told his master everything. Phileas Fogg was not happy.

'British law protects Indian sacred places,' he explained. 'I hope this does not bring us more trouble, but at least you are here on time.'

Passepartout understood how easily his master could lose his bet. He felt worried.

Yes, it was true, they were two days early, but anything could still happen.

---

5. **amazement** : the feeling you have when you're very surprised.

Three days later the train stopped at a small village. Passepartout heard the train driver shout, 'Everyone must get off. The railway line ends here!' The railway line from Bombay to Calcutta was not yet finished. The passengers had to travel to the next station, Allahabad, on their own. People who often travelled between the two towns were quick to find a way to continue their journey. Among the different types of transport there were little carts pulled by cows and ponies, and rickshaws pulled by bicycles or the men from the village.

Passepartout was worried about how to get to the next station, but Phileas Fogg immediately found a man with an elephant and after a short discussion, the man sold him the elephant for a very high price, and the two travellers were soon on their way to the next station with a guide and the elephant, called Kiouni.

At about nine o'clock that night our adventurers came to a big forest of palm trees where they had to stop to let Kiouni rest and eat the leaves from the trees. For a few days they slept in huts in the middle of the jungle. Sometimes they heard the cries of the monkeys or saw the footprints [6] of tigers. Their journey was going well until the elephant suddenly stopped.

6. **footprint** : a mark in the shape of a foot that a person or an animal makes in or on a surface.

The text and **beyond**

PET ❶ Comprehension check

Choose the correct answer — A, B, C or D.

1   Inspector Fix was waiting for
   A ☐ the warrant for Phileas Fogg's arrest.
   B ☐ the *Mongolia* to leave Brindisi.
   C ☐ Passepartout to walk past him.
   D ☐ Phileas Fogg to get on another ship.

2   What did Passeparout tell Inspector Fix?
   A ☐ His master was a very unusual man.
   B ☐ His master was going to Bombay because he is rich.
   C ☐ He did not have time to speak to strangers.
   D ☐ His master planned to travel the world in eighty days.

3   What did Passepartout buy in Bombay?
   A ☐ He bought sunglasses and two white shirts.
   B ☐ He bought sunhats, shirts and a pair of sandals.
   C ☐ He bought a pair of sandals and two train tickets.
   D ☐ He bought some gold jewellery.

4   The priests at the pagoda were angry because
   A ☐ it was against the rules to hit a priest.
   B ☐ Passepartout pushed one of them away.
   C ☐ they thought Passepartout was not showing respect.
   D ☐ Passepartout was wearing the wrong type of shoes.

5   The train stopped because
   A ☐ it always stopped for one night.
   B ☐ it was quicker to travel by cart or rickshaw.
   C ☐ it was dangerous to travel further.
   D ☐ the railway line wasn't finished.

6   How did Fogg and Passsepartout get to the next stop?
   A ☐ They walked through the jungle with the guide.
   B ☐ They bought an elephant to take them with a guide.
   C ☐ The man who sold them the elephant took them.
   D ☐ They took a rickshaw and walked through the jungle.

PET ② Signs

Look at the text in each question. What does it say? Choose the correct letter, A, B, or C.

1

> From: det-reed@met-police.co,uk
> To: insp-smith@met-police.co.uk
>
> I'm pleased you can now read your email. Can you send your report to the British Consulate this week?
>
> Regards,
> Detective Reed

A ☐ Inspector Smith is still having problems opening his email.

B ☐ Detective Reed must send a report to the British Consulate.

C ☐ Inspector Smith must give a report to the British Consulate.

2

> Incredible Indian Tours
> Tourist Booking Centre
> Elephant Tours
> (2 days) every Monday

A ☐ You can book tours every Monday.

B ☐ There are elephant tours every week.

C ☐ You go to a hotel on the elephant tour.

3

> JAIN TEMPLE:
> MALEBAR HILL
> VISITORS ARE
> WELCOME BETWEEN
> 10 AM AND 3 PM
> PLEASE REMOVE
> SHOES

A ☐ You can visit the temple at 12 midday.

B ☐ The visit to the temple takes five hours.

C ☐ Visitors not wearing shoes are welcome any time.

4

> **Warning:**
> Thieves operate at this station.
> Passengers should keep bags
> with them at all times.

A ☐ Report thieves to the police at the station.

B ☐ Hold your bag even if someone tries to steal it.

C ☐ Watch your bags. A thief could steal your luggage.

3 A new type of tourism: the reality tour

Read the text and fill in the gaps with the words in the box. Do you agree with this type of tourism?

> reason   tour   temples   visit   guides
> films   interested   tourism

At one time sightseeing in Mumbai probably included a tour to see important city buildings **(1)** ...., gardens or maybe Indian film celebrities, but in the past few years Western tourists are less interested in the palaces of the rich and more **(2)** .... in some of the world's poorest and most crowded areas: the 'slums'. Popular **(3)** .... like 'Slumdog Millionaire', the story of an Indian boy from Dharavi, a poor area of Mumbai, who wins the £20,000,000 rupees on the TV show 'Who wants to be a Millionaire?' are part of the **(4)** .... for the growing interest in these areas. Tourists are curious, not only about local customs, but also about how people live there. Not everyone wants to see this type of **(5)** .... They don't want their ordinary lives to become a tourist attraction or for people to think of them as 'poor'. On the other hand, they can see that this type of tourism can be a good thing. One **(6)** .... organiser, for example, gives the money they make to schools and they use people who live there as **(7)** .... The tourists visit the many small industries that make the things we buy every day. Many tourists will say they have learnt something new about life and about happiness from their **(8)** ....

Before you read

1 Listening

track 07

Listen to the first part of Chapter Four. Underline the incorrect facts in these sentences and correct them.

1 The people of the village are celebrating a wedding. ....

2 The wife of the prince is from a poor family. ....

3 Phileas Fogg says there is no time to save Mrs Aouda. ....

4 Passepartout ran towards the fire to save Mrs Aouda. ....

2 Vocabulary

A Use a dictionary to find the words in the word box. Which of these words can you find in the pictures? (There is more than one word for each picture).

bonfire   flames   storm   sunrise   clouds   town

B Complete the puzzle with the words to spell the name (in the shaded squares) of a tradition Phileas Fogg is both curious and surprised about in Chapter Four.

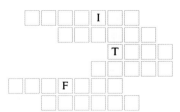

When our adventurers rescue a woman from certain death

They stopped near a village, where they heard the sound of strange musical instruments. Their guide went to discover what was happening and he was soon back with the news. The people of the village, their guide said, were celebrating a local tradition called *suttee*. Our travellers immediately wanted to know more about it. The guide told them that when a woman's husband dies, his wife must die with him and they burn her alive in a big fire.

track 07

'The name of the woman is Mrs Aouda,' he told them, 'and she's very beautiful. She is from a rich family. Her father was a businessman. Her parents sent her to a European school in Bombay. She learnt European languages there, and she has European ways. Her parents died and she had to marry an old prince. She became a widow after only three months. They are taking her to the pagoda tonight. They are going to burn her alive tomorrow, at sunrise.'

'My goodness! How terrible! Do such traditions still exist?' asked Phileas Fogg.

'Poor woman!' whispered Passepartout.

'We can still save her,' said Phileas Fogg. 'We are a few hours ahead of time.'

'Yes, but sir, if we save this woman, they'll try and kill us!' said their guide.

'I can only speak for myself, but I am prepared to take that risk,' replied Phileas Fogg.

'Me, too!' said Passepartout. When they arrived they started to plan the rescue. Unfortunately there were guards all around the pagoda and so they decided that it was too dangerous to do anything. They were about to leave, when Passepartout said that maybe he had an idea. When the sun came up the next day, the crowd [1] arrived to see the bonfire ready for the sacrifice. [2] Our travellers disappeared among the people. They saw the dead prince and his young wife through the smoke. Phileas Fogg prepared himself to run towards the fire in a final effort to save

---

1. **crowd** : large group of people.
2. **sacrifice** : the killing of an animal or person in a special religious ceremony as an offering to a god.

Mrs Aouda, when suddenly a terrified cry came from the crowd. Her husband was not dead! He stood up in the flames, took his wife in his hands and ran in the opposite direction to the crowd. It was not difficult to imagine Phileas Fogg's surprise when he later discovered that the woman's 'husband' was Passepartout. A few moments later our heroes disappeared into the forest with their new travelling companion, followed by the angry guards.

Mrs Aouda slowly started to wake up when they reached the station at Allahabad. Phileas Fogg thanked his guide for his loyalty and gave him the elephant. For a young guide an elephant like Kiouni was a big present. He could make a lot more money now that he had his own elephant. He was very happy and continued to thank Mr Fogg and the others until they left.

On the train to Calcutta, Phileas Fogg and Passepartout told Mrs Aouda all about their adventure. Mrs Aouda couldn't believe it: these men risked their lives — for her! She remembered there was a cousin with a business in Hong Kong and hoped he was still there. Phileas Fogg promised not to leave her until she was safe.

At seven o'clock they arrived in Calcutta. The ship for Hong Kong did not leave until twelve o'clock midday. Fortunately they were still on time — or, so they thought.When Phileas Fogg and Passepartout arrived with Mrs Aouda at the port of Calcutta, an official stopped them.

'I'm afraid you have to stay here. You must speak to the police,' he said.

'If it is about this lady, then you should know that she is here to escape the practice of *suttee*, which is against the law in this country and she has a right to leave.'

'A lady? No, no. The complaint is from a priest in Bombay.'

But how did the police in Calcutta know about Passepartout's adventures in Bombay?

The answer was simple. Inspector Fix. How did he get to Calcutta? How did he know where they were going? That remains a mystery, but one thing was sure: Inspector Fix was determined to stop Phileas Fogg and his visit to the police while in Bombay gave him the very reason he needed.

The official on the other hand, did not want to keep them. It was much easier he said for Phileas Fogg to pay the fine for his manservant's behaviour. Despite Passepartout's protests, with only a few hours to go, Phileas Fogg paid the fine. For Fix, all seemed lost. The warrant was not in Calcutta as he thought, but still, he had a plan. He just had to wait a little longer.

On the ship, all was well, until a bad storm near Singapore meant the ship was a day late. If the ship for Yokahama left on time, it was too late. His master could lose everything.

At the port in Hong Kong, Phileas Fogg went straight to the Captain.

'When does the next boat to Yokohama leave?' Fogg asked. 'Tomorrow morning,' he replied.

'Didn't it leave this morning?'

'No, they had to repair it, so it's not leaving until tomorrow.'

Passepartout was very happy to hear this good news.

Phileas Fogg simply wrote how late they were in his diary.

6 November – minus 24 hours

The text and **beyond**

1 Question words

Look at the sentences below. First put the words in the answers in the correct order. Then choose the best question word (A-E) for each sentence (1-5).

A Where **B** Who **C** What **D** When **E** Why

1 .... did the guide tell them about 'suttee'?

a/must/sacrifice/his wife/for/husband/her life/When/dies/him/woman's

2 .... was Mrs Aouda?

widow/an/The/prince/of/old

3 .... did they rescue Mrs Aouda?

The/when/next/came up/sun/the/day.

4 .... did Fogg give the guide an elephant?

his/wanted/He/to thank/service/for/him.

5 .... did the police stop Fogg and Passepartout?

stopped/port/They/at/them/the.

2 Characters

Decide if these statements about Mrs Aouda are true (T) or false (F). Correct the false ones.

		T	F
1	She was the daughter of an Indian prince.	☐	☐
2	She received a good education.	☐	☐
3	She has European ways.	☐	☐
4	She was clever, but she was not beautiful.	☐	☐
5	Both her parents died when she was a little girl.	☐	☐
6	She did not want to marry her husband.	☐	☐
7	She became a widow after a year.	☐	☐
8	She has an uncle with a business in Hong Kong.	☐	☐

3 Discussion

Discuss the questions below.

1   In Chapter Four, the practice of *suttee* meant the death of the wife. After many years, this practice stopped.
    Do you know of any traditions that are no longer practiced but the celebration remains? In your opinion, is it possible to 'celebrate' such things?

2   Passepartout is in trouble because he did not understand the importance of local customs. Make a list of customs you know about in other countries that are different to you own.
    Have you ever been in trouble for not knowing about a local custom?

4 Past Simple

Complete these sentences about Chapter Four with the verbs from the box, in the past tense. When you finish, put the sentences in the order you read about them in the story.

> give   hear   leave   run   send
> stand   take   tell   write

A ☐ They .... her to the pagoda to burn her alive with the body of her husband.

B ☐ Much to their surprise Passepartout .... up in the flames and .... in the opposite direction.

C ☐ The ship for Hong Kong .... twenty-four hours later.

D ☐ After their long journey Phileas Fogg thanked their guide and .... him the elephant.

E ☐ The guide .... them everything about the local tradition.

F ☐ They .... the sound of strange musical instruments.

G ☐ Phileas Fogg .... how late they were in his diary.

H ☐ Her parents .... her to a European school in Bombay.

PET **⑤** **Listening**

You will hear a radio announcement about a local event. Fill in the missing information in the numbered spaces.

Event: Notting Hill (**1**) .... in month of
(**2**) .... in the city of (**3**) ....
Things to do: watch the (**4**) ...., listen to
(**5**) ...., eat food.
Number to call about favourite things to do: (**6**) ....
Best way to get there: Underground (some stations are
(**7**) .... in the Notting Hill Gate area); Bus number
(**8**) .... from Victoria (**9**) .... or walk.

T: GRADE 5

⑥ **Speaking: Festivals**

We've just read about people celebrating this unusual tradition called *suttee*. Work with a partner and ask each other the following questions.

1 Have you ever been to a festival? If so, which one?

2 Can you describe it?

3 Can you name some of the most popular festivals in your country?

4 Which one do you like the most and why?

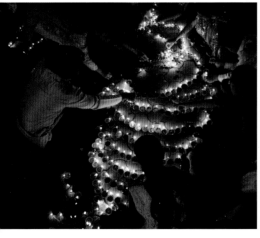

7 LOGS

On the journey, Phileas Fogg records times and dates of their arrivals and depatures. On a ship you can find this type of information in the ship's log. Read the information about logs below and answer the questions.

BLOG   CRÉER UN BLOG   Inscription   Annuaire   Visite guidée   Forum   Aide

Go Travel Blog

From ship's log to web log

I've been doing some research on logs for my travel blog and this is what I have found.

A logbook was a book used for writing down distances and times. Ships used to keep logbooks so the captain could see how far and how fast the ship travelled. Today ship's logs have other types of information, for example, details about the weather and important events, as well as the ports the ship visits. I also found out in my English class that the word log is not just used for ships. It can mean a place where we record information, for example, a way of keeping a record of problems we find with a machine. The person who keeps the log should add something every day or when anything new or important happens (we call this 'updating' the log).

One of the most recent changes in the use of 'logs' is in a popular activity called 'blogging'. A blog (web log) is a website anyone can create – like this one! People use blogs to write down their thoughts, ideas and opinions; it is like a diary (Americans call it a journal) on the Internet. If you have never used a blog, now is the time to start. I hope you enjoy reading my travel blog. I just need to remember to update it!

1 What type of information is this?

2 What information was kept in a ship's log?

3 How often should you write in a log?

4 Why is a blog like a diary?

Imbarco di Colombo, Pelagio Palagi, 1826-28.

Great journeys
around the world

Travelling East

For Europeans living in the Middle Ages, the most famous person to travel to the East was Marco Polo. Marco's father and uncle visited China when he was a little boy. They met Kublai Khan who ruled [1] all of China at that time. They promised to return and some years later they did; this time Marco came with them. Marco told the story of his adventures in the new worlds he visited between 1271 and 1291 and the time he spent in the court of Kublai Khan in *The Travels of Marco Polo*.

Later, ships from Europe travelled east around Africa and India, but very few travelled west.

---

1. **ruled** : was in control.

Travelling West

Christopher Columbus (born in Genoa, Italy) knew the world was round but he thought it was smaller. He thought by travelling west, a ship could be in China in less time than by travelling east. This was because nobody in Europe knew of the land we now call America or the Pacific. In 1472 he was finally given the money for his journey by the King and Queen of Spain. He was so sure his ship was sailing to the 'Indies' (modern-day Indonesia) that when he arrived at the Caribbean Islands he called the people there 'Indians'. When he described the places he visited, people asked if this really was the East or a new land.

The first person to cross the Pacific was the Portuguese explorer [2] Ferdinand Magellan. Magellan found a route to the Pacific through Panama. Ferdinand Magellan died in the Philippines. His men took one ship back to Spain. At the end of the journey only eighteen people were alive; they were the first people to sail around the world.

From South Pole [3] to North Pole

In the late 1700s, a large area of the South Pacific and Antarctica remained a mystery. In his second journey to find a land called 'the southern continent', the great explorer Captain Cook sailed around a strange island where it was always cloudy and said noone could live there. This island of ice was not the 'southern continent', it was Antarctica.

Captain Scott travelled across more of the Antarctic than anyone of that time, in some of the coldest temperatures on earth. He hoped to

2. **explorer**: an object used for travelling over snow.
3. **Pole**: the north or south end of the earth.

Landing of Columbus, John Vanderlyn, 1775-1852.

be the first man to arrive at the South Pole, but in 1912, the Norwegian Roald Amundsen was also in Antarctica and arrived at the South Pole first. Scott and his men died on the return journey, not knowing that they were close to food and water. Roald Amundsen was the first man to travel to both the North and South Poles, but it wasn't until 1979 that anyone travelled around the world from 'Pole to Pole' in one journey.

An expedition, [4] led by the adventurer Sir Ranulph Fiennes, went across the Sahara and Africa, then by ship to Antarctica. They travelled to the South Pole on foot, before sailing across the Pacific all the way to Canada. They arrived at the North Pole with the help of sledges [5] before returning to London after their amazing journey in some of the most difficult climates known to man.

4. **expedition**: an organised journey for science, research, etc.
5. **sledge**: an object used for travelling over snow.

Steve Seaton, Peter James, Sara Odell and Sir Ranulph Fiennes of Team Hi-Tec
during the glacier hike stage twenty of the 1999 Eco Challenge in Patagonia.

Different ways to travel

In 1884, Thomas Stevens left San Francisco to travel across America
but he decided to continue his journey and became the first man to
travel the world on a bike. Since then, people have used many
different types of transport: by helicopter, in a balloon and even by
walking and running.

The Jules Verne Trophy is given to the person who goes around the
world in any type of yacht in the fastest time starting in France and
finishing in England. In 2010 a French man completed the journey in
48 days, 7 hours, 44 minutes and 52 seconds.

One man is travelling around the world without leaving England. A
computer is showing his journey while he tries to row [6] over 40,000
kilometres in three years on a rowing machine. If he completes it, he
is hoping to receive thousands of pounds; he will give the money to
charity.

6. **row**: move through water with oars.

1 Comprehension check

Number the events (A-F) in the order they happen.

A [ ] Thomas Stevens rides around the world on a bicycle.

B [ ] Magellan's men are the first people to go around the world.

C [ ] Marco Polo meets the Kublai Khan.

D [ ] Captain Cook sails around Antarctica.

E [ ] Sir Ranulph Fiennes travels from Pole to Pole.

F [ ] Columbus visits 'the Americas' for the first time.

 INTERNET PROJECT

Find out more about famous explorers and adventurers.
Connect to the Internet and go to www.blackcat-cideb.com.
Click on the title of the book and on the Internet project link.

▶ In groups choose a different explorer/adventurer. Make a poster showing this person's life and/or travels.

8 **Your trip around the world**

Connect to the Internet and go to www.blackcat-cideb.com. Click on the title of the book and on the Internet project link.

On your around the world trip you go to a traditional festival.

Use the Internet to find out about traditional festivals in:

a India
b Hong Kong

You decide to start a travel 'blog'. You want to write what you did like a diary, but also about your thoughts and feelings. Write a report on your first day at a festival. There is an example of the first few lines below.

We went to the Holi festival today. If you go to India around the end of March, you must go to the North of India and see this. There were people throwing coloured water everywhere so my camera is yellow!

Luckily, James brought his camera so I've posted his photos.

Think about other ways of communicating with your friends at home. Write them here.

Before you read

1 Listening

Listen to the first part of Chapter Five. For each question tick (✓) A, B or C.

1 Where are they?

A ☐ in Singapore B ☐ in Hong Kong C ☐ in Yokohama

2 Inspector Fix is waiting for

A ☐ the Hong Kong police. B ☐ Phileas Fogg.

C ☐ the arrest warrant.

3 Passepartout goes to

A ☐ a bar. B ☐ a hotel. C ☐ to see Mr Fogg.

2 What happens next?

Tick (✓) the option you think does not happen next.

A ☐ Passepartout hits Inspector Fix. B ☐ Passepartout falls asleep.

C ☐ Passeparout disappears.

3 Word game

Write a word in the box to describe the person in the picture. Then match the people (1-3) to the things they can do (A-C).

A ☐ keep objects in the air B ☐ pull a funny face

C ☐ do gymnastics

Put the shaded letters above in the correct order to complete this sentence.

Passepartout finds a new _ _ _

CHAPTER FIVE

When Passepartout becomes an acrobat

Hong Kong was the last country they travelled to under British law. Inspector Fix was on the ship and this was his final opportunity to get an arrest warrant for Phileas Fogg. It was not difficult to imagine his anger when he discovered that the British officials in Hong Kong knew nothing about the warrant and told him he had to wait again. His only chance to arrest Phileas Fogg was to keep him in Hong Kong, but how?

While he was thinking about this, Fix recognised Passepartout walking down the steps of the *Carnatic*, the ship which was taking them to Yokohama. An excited Inspector Fix ran towards him and shook his hand.

'Fix. We met in Egypt, in the Suez', he said a little out of breath. [1] Passepartout was surprised by this sudden introduction.

'I'm sorry, Mr Fix, but I need to go back to the hotel to tell my master that this ship isn't leaving tomorrow morning. The repairs are complete; it's going to leave tonight.'

'Of course, I understand,' Inspector Fix began. 'But, it's still early, I know a nice place near the hotel. I'm sure we can stop for a drink. It's difficult to find a good friend to talk to after such a long journey away from home.'

'Well, just one drink...' said Passepartout. He remembered from the last journey that Mr Fix was good company.

end

When they were inside the bar, Inspector Fix gave the barman some money and asked him to make a very strong drink for Passepartout. Less than an hour later Passepartout was drunk and fell asleep at the table.

The detective's plan was complete.

'Phileas Fogg won't know about the ship until it's too late,' he smiled. 'And now Mr Fogg, I just need to wait until I have the warrant for your arrest, and it won't be long now, you can be certain of that!'

The next day, when Phileas Fogg woke up, he saw Passepartout was not in his room. He packed the suitcases himself and paid the hotel bill. Then he went straight to the port. Mrs Aouda came with him; after spending the day before looking for her cousin she was told that he didn't live in Hong Kong any more, in fact, his neighbour thought he was in Europe.

'You must come with us, madam. I will not hear 'no' for an

---

1. **out of breath** : having difficulty to take air into his chest and speak normally.

answer,' said Phileas Fogg when he heard the news. Mrs Aouda knew her answer. To look for a man she didn't know was difficult, but returning to India was impossible.

At the port Phileas Fogg looked for his manservant. Passepartout was not at the port and their ship, the *Carnatic* was not there either. Mrs Aouda was very worried but Phileas Fogg did not seem worried about the departure of his ship or Passepartout, instead he went to speak to the captains of the different boats in the port.

Inspector Fix, who was waiting for Phileas Fogg, followed behind. What was Phileas Fogg's new plan? Fix saw him pay the captain of a small ship, who agreed to leave immediately. 'Oh no!' thought Inspector Fix. 'He always thinks of something. That thief! He can't escape! Not now.'

Fix walked up and down near the ship, until Phileas Fogg noticed the poor man.

'Are you looking for a ship, too, my good man?' he asked. 'We're going to Japan. If that's the direction you're going in, you're welcome to come with us.'

'Thank you, sir. That's very kind of you. My ship left early and I am in a terrible situation, in fact, I was just thinking about how I could find another ship to take me to Yokohama,' replied Inspector Fix. He was amazed at his good luck but he was worried that things were not going exactly as he planned. He was determined to catch his thief. 'I'll have to follow Fogg around the world if that's the only way I can catch him,' he thought.

But where was our friend Passepartout?

Passepartout woke up in the bar a few hours later. His head hurt and he could not remember anything, except that he was in

a bar and ... the ship! The *Carnatic* was leaving that evening. He looked at his watch. 'Oh no!' he thought. 'I'm late. Mr Fogg will be at the port now.' He ran to the boat and got on. He looked everywhere for Mr Fogg and Mrs Aouda, but he could not see them. Then he started to remember his afternoon. 'But of course!' he thought. 'Mr Fogg doesn't know. How could he? I was still in that bar with Mr Fix, and then I ...'

It was too late. The ship was already sailing towards Yokohama.

He felt very bad. This was terrible. His master was losing his bet and he did not have a penny in his pocket!

When he arrived in Yokohama, he walked around the streets, trying to decide what to do. Soon he felt hungry and he decided to sell his elegant European jacket and buy an old Japanese one, but the money was not enough. He needed money to eat and to sleep, and, above all, to return home. Just when he thought the situation was hopeless he saw an advertisement for a circus.

> Don't miss Batulcar's Circus.
> The last show before the circus moves to America.
> Acrobats, clowns, lions, tigers and much more!

'What luck!' thought Passepartout. 'I'll go to the circus owner. If he lets me go with them, I can go to America, and from there to England.'

'So, you say you're from Paris?' said Mr Batulcar, a big man with a bald [2] head and a moustache. He looked at Passepartout carefully.

'Yes, a true Parisian, from Paris,' replied Passepartout.

'Well, you know how to make funny faces then,' said Mr

2. **bald** : head with little or no hair.

Batulcar smiling from the corners of his moustache. 'You can start as a clown,' said Mr Batulcar.

That night an acrobat was ill, and Passepartout had to take his place as part of a human pyramid. Passepartout was at the bottom of the pyramid and he had to carry the weight of several men on his shoulders. The audience shouted out loudly, the drums sounded like a thunderstorm, and then... the acrobats in the pyramid fell to the floor like a pack of cards. Passepartout run towards someone in the audience, someone he was very happy to see. It was his old employer, Phileas Fogg. But how did he get there?

When Fogg and Mrs Aouda arrived in Yokohama, they spoke to the captain of the *Carnatic* and found Passepartout's name among the list of passengers, but his ticket only took him to Yokohama. Phileas Fogg and Mrs Aouda wanted to find Passepartout before the ship left for its next destination — San Francisco. Inspector Fix stayed with them, telling them he must return to Europe via America on business and why not with friends?

They looked all over the city but they could not find Passepartout anywhere. Then Phileas Fogg saw the sign for the circus. 'Interesting,' he said to himself. 'Let's go and see the circus before we leave town,' he told them, 'I believe it's very good.' Phileas Fogg did not see his manservant among the acrobats, but his manservant certainly saw him, and he left everyone else in the human pyramid like a group of arms and legs on the floor.

They had no time to say sorry to a very angry Mr Batulcar. The *Carnatic* was leaving for America.

The text and **beyond**

PET ① Comprehension check

Choose the correct answer — A, B, C or D.

1 Inspector Fix is not happy because
   A ☐ the *Carnatic* was not leaving for 24 hours.
   B ☐ the officials knew nothing about the warrant.
   C ☐ Passepartout did not have time to speak to him.
   D ☐ Mr Fogg did not plan to stay in Hong Kong.

2 Passepartout and Inspector Fix
   A ☐ were not very interested in Mr Fogg's journey.
   B ☐ were good friends who enjoyed talking.
   C ☐ wanted to help each other.
   D ☐ had different reasons for going to the bar.

3 Inspector Fix's planned to
   A ☐ make sure Passepartout wasn't on the ship.
   B ☐ stop Passepartout from leaving Hong Kong.
   C ☐ stop Fogg from leaving himself.
   D ☐ make sure Fogg didn't find out about the ship.

4 Who sailed on the *Carnatic* to Yokohama?
   A ☐ Inspector Fix sailed on the *Carnatic* to Yokohama.
   B ☐ Mr Fogg and Mrs Aouda went to Yokohama.
   C ☐ Passepartout sailed on the *Carnatic* to Yokohama.
   D ☐ Noone went on the *Carnatic* to Yokohama.

5 Why did Passeparout get a job with the circus?
   A ☐ He knew Mrs Aouda liked the circus.
   B ☐ He lost the bag with the money to Inspector Fix.
   C ☐ He could return to Europe and he had the experience.
   D ☐ He preferred the circus to being a manservant.

6 Inspector Fix said that he was travelling to America
   A ☐ because he was on his way to Europe on business.
   B ☐ because he was a good friend of Passepartout's.
   C ☐ because he wanted to tell the American police about Fogg.
   D ☐ because he hoped to find out more about Fogg.

2 Telegrams

In 1872 the best way to send a message to another part of the world was by telegram but you had to pay for each word so they were very short and sometimes missed words and punctuation. Match the telegram to the people and places they are sent to. Add commas, full stops and the missing words to the telegram messages.

A ☐ Scotland Yard, London   B ☐ British Embassy, Hong Kong

C ☐ British Consulate, Calcutta   D ☐ Indian Railway Police

> **1** Criminal seen 7 o'clock Brindisi port please send warrant for arrest to British Consul Suez
>
> **2** Travellers wanted for questioning in Bombay French manservant curly hair brown bag with master tall English brown moustache arrival 1600 hours
>
> **3** No warrant for arrest arrived criminal paid fine and is leaving send warrant to Hong Kong immediately.
>
> **4** Do not let Fogg continue to Japan Warrant went to wrong office Bombay Head Office will send this week

3 SMS

Read the information about SMS. Try to re-write the text messages below using full sentences.

Like telegrams, text messages are brief written messages. When we use a mobile phone, we use SMS (short message service) or 'text messaging'. In text messages we can use numbers to make words, for example: gr8! ('Eight' sounds like the end of the word 'great' so gr + 8 = great.) There are also abbreviations, e.g. LOL = laugh out loud) and we can miss letters out or write things as they sound. (K =OK, gud = good etc.)

What r u doin 2nite?
Going 4 pizza.
Want 2 meet @ 8?
X

K ! I luv pizza
C u l8R!

....

....

4 **Crack the code!**

Inspector Fix has sent a telegram but he has written part of it in code. A police officer at the office has made some notes. Can you guess the message?

> To: Calcutta British Administration
>
> 6 November 1872
>
> PLEASE SEND
>
> PTC HBMMBRP ZFM PTC BMMCNP FZ ZFUU PF
>
> HONG KONG
>
> Inspector Fix

In code - A = B, E = C, I = D, O = F, U = G, F = Z, H = W, M = R, T = P ..... !

5 **Characters: motivation**

Sometimes we ask: why does this person behave this way? This is their 'motivation'. Sometimes it is clear, but sometimes there are other reasons, too. Look at the characters of Inspector Fix and Passepartout and answer these questions using what you know from the story and your own ideas.

Inspector Fix: why does Inspector Fix want to be friends with Passepartout and then Mrs Aouda and Mr Fogg? Do you think Inspector Fix needs friends?

Passepartout: is Passepartout happy to be on his own? Why is he not angry with Inspector Fix when they meet again?

6 **Your Around the World Trip**

As part of your Around the World Trip you are either a) going to find paid work or b) work without being paid and do something you enjoy that helps people or animals (volunteering).

track 10

Listening

You will hear an interview about working abroad. Fill in the missing information in the numbered spaces.

WORKING ABROAD

Age group the organisation speaks to : 18 — (**1**) ....

Jobs people do: working in (**2**) ....,

(**3**) .... English, working with animals, e.g.
(**4**) ....

(Working with animals you sometimes pay for travel, accommodation, (**5**) ....)

Places - Paid work: Japan, New Zealand, (**6**) ....

Volunteer work:, Africa, Asia (**7**) ....

Website address: (**8**) ....

Discuss these questions with another student or write down the answers yourself.

- Are you going to look for paid work or to volunteer?
- Which country are you going to look for work in?
- What do you want to do there?

PET⑧ Writing

This is part of an email you receive from an English friend. Reply to the email telling her about your plans for working abroad. (100 words)

> Hi!
> How's your trip? Rachel told me you are looking for work. What are you going to do? I want to hear about all about your plans. Speak soon. We miss you!
>
> Christina

Before you read

1 Reading pictures

Find these words in the picture on page 71.

carriage   snow   buffalo   bridge

passengers   coal   driver   engine   rail tracks

 INTERNET PROJECT

Connect to the Internet and go to www.blackcat-cideb.com. Click on the title of the book and on the Internet project link.

Read this information about the western states of America. From the information you have, write the name of a place under the photos.

Come and visit America's wild west!

The West of America is a land of contrasts: from the high Rocky Mountains and deep ravines of Colorado to the low Death Valley, the driest desert in North America, and on to the Pacific Ocean and the tropical climate of the Gulf of Mexico. If you want space, the miles of plains of the Mid-Western states will give you all the peace and quiet you need; or if you prefer the city lights, then some of the world's most famous cities: San Francisco, Los Angeles, and the casinos of Las Vegas are all here in the Western states of America.

Which section do you need to look at to find:

▶ information on visas?

▶ things to do?

▶ places to stay?

CHAPTER **SIX**

When our friends risk their lives

They were now sailing across the Pacific on the *Carnatic* in the direction of San Francisco. On the journey Passepartout began to ask questions about Mr Fix. There was something strange about the fact he was doing the exact same journey as them. Maybe he was a spy from the Reform Club. Of course! Why didn't he think of it before? He wanted to make sure Phileas Fogg was really completing his journey. But did he want him to get drunk in that bar? Was he trying to stop his master from winning his bet?

'I'm going for a walk. I believe Mrs Aouda will join me. I'll see you in the morning at seven-fifteen,' said Phileas Fogg, interrupting his manservant's thoughts. Passepartout decided not to tell Fogg. He was an honest man for the Reform Club to

question him in that way! He decided instead to watch Mr Fix carefully to make sure nothing else happened.

When they arrived in San Francisco, Phileas Fogg made a note in his diary:

> Tuesday – 2 hours ahead.
> Wednesday – 3 hours behind.
> Thursday – arrived in San Francisco on time.

As there was time until the evening to catch the train, Mrs Aouda suggested they all go into the city, including their new friend, Mr Fix. It became clear that Mrs Aouda was very close to Phileas Fogg. He, on the other hand, did not seem to notice the beautiful lady by his side but he did everything to make her happy and comfortable.

There were people from all over the world in San Francisco and Passepartout thought it was a beautiful place with its hills and views.

On their way back they went past a square where a large crowd of people were waving banners [1] and shouting. On the banners were the names of two political parties. They all waited with interest to see what the protest was about. Soon the shouts became angry and they only moved a few steps before they were in the middle of a fight. A man from one group hit a man from the other over the head with his banner. Phileas Fogg took Mrs Aouda by the arm, Passepartout pushed past as best as he could but just at that moment a red-faced man with white hair and a thick moustache tried to hit Phileas Fogg in the face.

'Is this the American way?' Fogg asked the man coldly.

---

1. **banners** : long strips of cloth with something written on them, usually carried during a protest.

'Better than the English way, coward!' replied the man.

Inspector Fix arrived. 'Do not call an Englishman a coward!' he shouted.

'If you want to end this now with a duel, we can.' The man replied.

'I'll happily defend my honour,' said Fogg. 'But it'll have to wait until another time.'

'I'll hold you to your word. My name is Colonel Proctor and I am going east to Fort Kearney tonight. If I should meet you in your travels, take a gun with you!' And with these words he turned angrily to another man while they escaped as quickly as possible.

Now they were in America, said Mr Fix, and buying a gun was not a bad idea. Phileas Fogg did not agree but Passepartout asked his master to stop at a shop to ask the price. 'I hear the train journeys here can be dangerous, too,' he added.

The same evening, at exactly six o'clock, our adventurers left San Francisco to travel to New York. The journey that once took six months to complete, now took seven days on the new Union Pacific Railroad that took the passengers from San Francisco in the west, to Omaha in the central state of Nebraska. From there Phileas Fogg hoped to continue to New York for the final part of their journey: crossing the Atlantic to England on 11 December.

After just one hour it started to snow heavily. At about nine o'clock the next morning, the train stopped. However, to their amazement it was not because of the snow but because hundreds of buffaloes were crossing the tracks in front of the train.

Passepartout became very impatient. 'I can't believe this!' he shouted. 'This country has a modern railway and the train must stop for buffaloes!'

The train driver told them he had no choice. The buffaloes could damage the engine. They had to wait until they moved across the tracks — three hours later!

As they went through the mountains in Wyoming, Phileas Fogg taught Mrs Aouda how to play cards. Mrs Aouda was a very patient learner, and they were soon so occupied with their games that they did not even seem to notice the deep ravines below them. Suddenly they almost hit the seat in front of them.

The train stopped and gave several loud whistles. [2] Passepartout got up to see what the problem was. He saw the driver talking to a man from the next station, a place called Medicine Bow.

'The station guard sent me to tell you that you can't go any further,' he said. 'The bridge across the ravine is not safe and it can't take the weight of the train. We have sent a telegram to Omaha, but it will be six hours before another train arrives.'

'We can't stay here all night. We'll die of cold in this snow!' shouted one of the passengers, hearing their conversation.

'Yes, but it'll take six hours to go on foot to the next station,' said the train driver's assistant.

'I think I have an idea,' said the train driver. 'We can get our train across the bridge, if we go fast enough.'

Passepartout was interested to hear more.

2. **whistles** : the sounds a train makes when steam comes out of it.

'If the train moves at its top speed, it won't be as heavy on the bridge,' he continued. 'And we can get across before the bridge breaks.'

Passepartout could not understand why the other passengers seemed to think this was a good idea.

'Isn't there a simpler solution, perhaps...?' he began to ask the driver.

The driver was not listening. 'No, no, this is the best solution we have,' he said.

'Yes, but maybe not the safest...'

Passepartout tried to explain that he had another idea.

'Maybe the passengers can go across the bridge on foot. Then the train could follow afterwards,' he said.

'No, the driver is right. If we go at top speed, we can get across the bridge,' said a strangely familiar voice behind them. The driver did as the passenger said. He blew the whistle and the train went back along the tracks about two kilometres. Then he blew the whistle again. The train moved faster and faster as it came closer to the bridge. Passepartout stopped and stared into the ravine, his heart was in his mouth; he knew the other passenger, then he was certain of it — it was the colonel.

In what seemed like minutes, or maybe it was only a few seconds, they were over on the other side, just in time to see the bridge fall into the deep ravine behind them.

The text and **beyond**

1 Comprehension check

Put the events in the correct order (1-7) to make a summary of Chapter Six. Then re-write the words underlined in order to make the last sentence (H) in the summary.

A ☐ A man tells Fogg he wants a duel. **It** is lucky that they have no guns.

B ☐ The train crosses over a deep **ravine**; they are too occupied to see it.

C ☐ **After** they leave San Francisco it starts to snow.

D ☐ Passepartout is worried for the passengers; the bridge could **falls into** the ravine.

E ☐ **The bridge** at Medicine Bow can't take the weight of the train.

F ☐ **They** wait 3 hours for buffalo to **go across** the tracks.

G ☐ Passepartout thinks Mr Fix is a member of **the** Reform Club.

H ....

2 Interview

Look at these notes made by a journalist after the bridge fell in the ravine. Write the questions in full using the past tense and match them to the answers. Write some questions yourself and practice the interview with another student.

1 Why/train/stop/at Medicine bow?

2 What/train driver/decide/to do?

3 How/you/feel?

A ☐ to go across the bridge at top speed.

B ☐ I felt scared.

C ☐ the train was too heavy for the bridge.

'If the train moves at its top speed … we can get across before the bridge breaks'

We can use the word 'can' in different ways.

- **To talk about something we are able to do:**
  I can speak English, but I can't speak German.

- **To be allowed/not allowed to do or to ask permission to do something:**
  You can't use your mobile phone here. / Can I open the window?

- **For a logical action (out of necessity):**
  If you're too cold, you can put on a jumper.

3 **Can**

Answer the questions.

1   In the example sentence above, describe 'can'.

  a ☐ able to   **b** ☐ permission to   **c** ☐ a logical, necessary action.

2   In the story of *Around the World in 80 days* find three things Passpartout or Phileas Fogg are able to do?

3   Think about Inspector Fix. Is there anything he is not allowed to do? (Use 'can' in your answer.)

track 12

4 **Listening**

Listen to the weather forecast and tick (✓) the weather you hear. Then match the words (1-6) to the pictures (A-F).

1 snow   **2** rain   **3** sun   **4** thunderstorm   **5** wind   **6** cloudy

A ☐   B ☐   C ☐

D ☐   E ☐   F ☐

PET⑤ Speaking

A man wants to travel across America. Talk about the different ways he could travel and say which you think is best.

 INTERNET PROJECT

Our protagonists spend a few hours in San Francisco.
Find out more about this city. Connect to the Internet and go to
www.blackcat-cideb.com. Click on the title of the book and on the
Internet project link.
In small groups organise your stay in San Francisco for the weekend,
choosing where to stay, what to do, where to eat and where to shop.

Before you read

1 Buffalo Bill

You will read the words in the box in Chapter Seven. Use a dictionary to find any words you don't know. Read this information about Buffalo Bill. Complete the text with the missing words.

> arrows   caught   leader   soldiers   horses   carry
> shooting   Indians   cowboys   forts

Buffalo Bill (real name William Cody) became famous all over America for the number of buffalo he (1) .... once, killing over 4,000 animals for their meat in eight months to bring to the (2) .... and workers on the railway. Bill (the short name for William) was born in Iowa in 1846. At the age of fourteen he went to look for gold, but he did not travel far. He became a rider with the *Pony Express*. The *Pony Express* rode their (3) .... for long distances to (4) .... messages and letters. Later he took food and other necessities to the soldiers in the (5) ....

In 1883 Bill started a show called Buffallo Bill's Wild West. Famous Cowboys and (6) .... played parts in the show. There were (7) .... competitions between (8) ...., stories of travelling on the trains, the fights with the Indians firing (9) .... at the passengers and the adventures of the brave (10) .... of the Sioux Indians, Sitting Bull. Although Bill himself fought the American Indians, they won his respect and some were his personal friends.

2 Reading pictures

Look at the pictures on pages 78, 80 and 81 and answer the questions below.

1   Who do you think the man in the picture on page 80 is?
2   What's happening in the picture on pages 82-83?
3   Who are the people riding the horses?
4   Why are the passengers using guns?
5   Who/what are they shooting at?

CHAPTER **SEVEN**

When Phileas Fogg comes to the rescue

Their journey continued across the high mountains of Colorado. The passengers were soon familiar with the delays; these only seemed to worry Passepartout. He continued to think about the Colonel. He did not want him to meet Phileas Fogg. When Mr Fix was asleep and Phileas Fogg got up to take a walk, Aouda spoke to Passepartout.

track 13

'I know that when you look worried it is usually about Mr Fogg, Passepartout. Please tell me what it is you are thinking.'

'There's a passenger on this train; we met him before in San Francisco and I wouldn't like Mr Fogg to meet him again.'

'The man who wanted to fight with him?'

'The very same man, madam. Maybe it wasn't a good idea to buy those guns.'

In three days and three nights, they travelled more than 2,200 kilometres. Phileas Fogg continued to play his games of cards with Mrs Aouda, while Inspector Fix slept like a baby, his head going up and down with the gentle movement of the train. Passepartout was not wrong in thinking that there could be more trouble. He was getting up to go to the window for some fresh air when a red-faced man with a moustache came towards him. The carriage door was open and before Passepartout could stop him, he walked straight in.

'Well, well! The Englishman and his pretty lady companion,' he said touching Mrs Aouda's face.

'I do not know what you want, sir, but this matter is between you and I alone,' said Fogg.

'I believe you agreed to duel a few days ago. Or are you a coward after all?'

'Very well, sir. I will defend my word as an Englishman but not in front of a lady.' He took a gun from the bag and before Passepartout or Aouda could say anything, he followed the colonel into the corridor towards the empty dinner carriage.

'Quickly! We must stop him!' said Mrs Aouda to Passepartout, but it was too late. Just at that moment there was a gunshot and then another. Inspector Fix woke up and asked a worried Mrs Aouda what was happening. (He did not want to come all the way to America to have a dead man on his hands and no reward!)

Passepartout was running towards the dinner carriage scared for his master's life but what he saw when he arrived surprised him more. Both Phileas Fogg and the colonel had guns in their

hands, but they were shooting them out of the window at a group of Sioux Indians who were attacking the train.

The Sioux rode their horses along both sides of the train. The passengers heard their battle cries before they arrived. Many of them knew the Sioux sometimes attacked the trains that travelled through these areas and prepared to defend themselves.

The horses of the Sioux moved faster and faster and several arrows flew towards the train carriages until finally the train slowed down. The Sioux jumped on the train. Their leader tied up the train driver and his assistant together and threw them off the train. Then the Sioux continued towards the other carriages.

'They're coming towards our carriage!' shouted Mrs Aouda.

Inspector Fix hit one of the men over the head with the back of his gun. Passepartout, in the corridor, fought bravely, using only his hands and his strength. 'We need to stop the train,' he said finally arriving at the same carriage as Phileas Fogg and the colonel.

'There are soldiers at Fort Kearney, a few miles away,' said the colonel. 'They are my regiment. I was on my way to join them. They will help us.'

Yes, but there's no driver', said Passepartout. 'We need to stop the train.'

Passepartout knew that there was only one way — to climb under the train. He opened the door and went under their carriage. Then with his great acrobatic strength he pulled himself along the bottom of the other carriages until he finally found the engine above his head. He separated the engine from the carriages and the train stopped. They were almost at Fort Kearney.

The colonel seemed to forget all about his fight with the Englishman now helping him fight the Sioux and he shook Fogg's hand as he got ready to leave the train. The soldiers at Fort Kearney heard the cries of the Sioux and the sound of the guns coming from the train. They quickly got on their horses and went to see what was happening. The Sioux were surprised to see that the soldiers were already shooting at them in the distance.

'Let's go!' said their leader. 'But first, take that man who is giving us so much trouble.'

The group rode away on their horses taking two passengers with them, and the brave young French man who was trying to save them.

When the battle was over Phileas Fogg could not find Passepartout.

'The Indians took him away,' said a passenger. 'Poor man! They'll kill him for sure!'

'I'll find him and bring him back, dead or alive,' replied Phileas Fogg.

Mrs Aouda looked into his eyes. He was her hero. He truly was a wonderful person.

'We can save these people, but I need help,' Phileas Fogg said to the soldiers.

He left the fort with thirty soldiers, and their horses, and followed the direction of the Sioux.

Mrs Aouda and Inspector Fix waited for him at Fort Kearney. It was incredibly cold and the wind was blowing hard. [1] Inspector Fix and Mrs Aouda sat in the uncomfortable station waiting room trying to keep warm. From time to time they looked outside at the snow. The darkness of the night started to become morning, but she still could not see Phileas Fogg.

Not long after the sun came up they heard the sound of guns in the distance. They stood up and looked out of the windows. But there was no battle, just the sound of celebrations.

A group of people, with Phileas Fogg in front, were coming on horses towards them. Passepartout and the two other passengers were sitting on the horses behind. They looked safe and well.

Mrs Aouda ran to meet them. Inspector Fix waited outside the station. 'Maybe he's not so clever after all,' he decided. 'But we must return to England soon and then I can arrest him.'

'You're all back safe! This is wonderful!' Mrs Aouda cried. Everyone, except Passepartout, looked happy.

'Yes, we're safe but Mr Fogg will probably lose his bet because

1. **blowing hard** : air is moving fast because of the wind.

of me,' said Passepartout. He left them to their celebrations and went to the station to find out about trains to New York.

'When's the next train to New York?' he asked.

'The next one leaves tonight,' was the reply.

'But we're already over twenty-four hours late. If the train leaves tonight, we'll be too late to get the boat!' Phileas Fogg came to rescue him and Passepartout felt very bad because he wanted to be a hero.

At that moment Inspector Fix returned with a man he was talking to outside the station.

'This man says he can take us to the station in Omaha in his sledge,' he said. 'We can take a train to New York from there.' The man had a strange sledge with sails. He explained to them that he often took passengers from one station to another in the winter, when the snow stopped the trains, and that with a good wind behind them, they could go a lot faster than the train.

Phileas Fogg agreed. They had no choice.

They all climbed onto the sledge. The sledge travelled very quickly across the icy, flat lands of the central states. The passengers were very cold, and with an icy wind blowing in their ears they did not speak for most of the journey. They were in Omaha in less than five hours. When they arrived, they thanked the man and Phileas Fogg paid him well.

Fortunately, they found a train to Chicago and then to New York immediately. They arrived in New York two days later at eleven o'clock on 11 December. They quickly went to the port, but the *China*, the ship taking them to Liverpool, was not there. Fogg did not look surprised. He looked at his watch. They were forty-five minutes late.

The text and **beyond**

PET **PET ❶ Comprehension check**

Choose the correct answer — A, B, C or D.

1 Passepartout is worried because
   A ☐ Mr Fix was trying to stop them.
   B ☐ the delays could cost Fogg his bet.
   C ☐ a man who wanted to fight Fogg was on the train.
   D ☐ they had only travelled 2,200 kilometres.

2 Who were Fogg and the Colonel shooting at?
   A ☐ They were shooting at each other.
   B ☐ They were shooting at the soldiers at Fort Kearney.
   C ☐ They were shooting at the Sioux.
   D ☐ They were shooting out of the window for no reason.

3 Passepartout separated the carriages by
   A ☐ climbing underneath them.
   B ☐ climbing on the roof.
   C ☐ going to the engine room.
   D ☐ opening the doors between the carriages.

4 What happened near Fort Kearney?
   A ☐ The Colonel started shooting at Fogg.
   B ☐ The soldiers started shooting at the Sioux.
   C ☐ The Sioux killed many passengers.
   D ☐ Passepartout couldn't stop the train.

5 The Sioux took Passepartout because
   A ☐ he tried to rescue two passengers.
   B ☐ he stopped the train before Fort Kearney.
   C ☐ they were going to ask Fogg for money.
   D ☐ they said he was giving them trouble.

6 Fogg travelled to New York
   A ☐ in a sledge with sails, stopping only once.
   B ☐ by sledge to Omaha and then by train.
   C ☐ by train to Omaha and then by sledge.
   D ☐ by sledge and then by ship.

2 **Characters**

A Look at the title of Chapter Seven. Who is really the hero? Passepartout or Phileas Fogg's? Choose the best answer in your opinion.

1 ☐ Phileas Fogg is the real hero because he does not get caught.
2 ☐ Passepartout is the real hero because he saves everyone's life, including Phileas Fogg's.
3 ☐ They are both heroes. Without each other, they could both be dead.

B Describe a character in Chapter Six but do not say his/her name. Let someone else guess who you are describing.

Ex.: This person... .

3 **Grammar duel**
In pairs find the missing words. If you know the answer to question 1 shout 'fire'. You must give the answer in 30 seconds. If it is wrong the other person can answer. Continue for all the questions. The person with the most correct answers wins the 'duel'. Write some more sentences with missing words and play again.

		P1	P2
1	The colonel came ..... their carriage.	☐	☐
2	Phileas Fogg agreed ..... a duel.	☐	☐
3	Passepartout was ..... late to stop them.	☐	☐
4	Fogg and the Colonel had guns in ..... hands.	☐	☐
5	Inspector could ..... get a reward if Fogg died.	☐	☐
6	The Colonel and Fogg ..... shooting at the Sioux.	☐	☐

4 **Listening and speaking**

track 14

Passepartout goes to the station to find out about trains. Listen to someone asking for information about trains and make a note of the reply. Use this information to create a similar dialogue of your own.

• Date and time of travel
• Destination
• Tickets and Prices

...the horses ... moved faster than the train.

A comparative can compare two things: *a car is faster than a bike,* or more that two things: *an aeroplane is faster than a bike or a car.* With three things, we could use a superlative adjective: *aeroplanes are the <u>fastest</u> (i.e. faster than a bike or a car).*

If we want to say that two things are/aren't the same we can use 'as ... as' or 'not as ... as': the sledge travelled *as fast as* the train or trains don't travel *as fast as* aeroplanes. (Note the word order. The adjective does not change).

5 **Comparative and superlative adjectives**

A **Complete the table below with the correct form of the comparative/superlative adjectives.**

Adjective	Comparative	Superlative
big	*bigger*	....
lovely	....	*The loveliest*
hot	*hotter*	....
intelligent	....	*The most intelligent*
bad	*worse*	....
good	....	....
sad	....	....
high	*higher*	....

B **Use the words below with** *as ... as* **or** *not as ... as* **or a comparative adjective to describe:**

1 cold: your country/England ....

2 big: your country/USA ....

3 exciting: travelling by balloon/travelling by train ....

4 hot: your country/India ....

5 long: River Nile / River Mississippi ....

6 high: Mount Everest / Mont Blanc ....

Compagnie Generale Transatlantique French Line, poster by Albert Sebille.

Passenger ships
and Transatlantic travel

Steam [1] and passenger ships

Until the mid 1800s, ships travelling from Europe to America were mostly carrying cargo. It was a dangerous journey; there could be storms, high winds and rough seas; but all this was to change. The industrial revolution meant that using the power of steam, engines became better and faster. It was possible to build a ship, from metal as well as wood, but there was still one problem: the ships did not leave or arrive at fixed times, which was not good for business. In 1839 Queen Victoria, offered the Canadian Sir Samuel Cunard the opportunity to begin a scheduled [2] mail service between the UK and

---

1. **steam:** water makes this when it is 100°C.
2. **scheduled:** always runs at the same time.

America. Mr Cunard saw another opportunity: the new class of rich businessman, travelling not only for work but for pleasure.

In 1840, a Cunard ship, the *Britannia*, took passengers from Southampton to New York and there were further services from Liverpool. On one ship there was a cow so passengers had fresh milk. As the ships became bigger and better, facilities included electricity and rooms for washing and by the 1870s the first and second class passengers had the same facilities as a good hotel. The luxury of the first class accommodation showed how beautiful a ship was, but luxury was not the only thing companies needed to offer. They could make good money, as before, by carrying emigrants (people looking to live and work abroad) in cheap accommodation at the bottom of the ship. Third class, also called 'steerage', was crowded and the passengers brought their own food. The speed a ship travelled was also important. A 'prize'[3] called the 'Blue Riband' was started for the fastest ship to cross the Atlantic. In reality, the winner did not receive anything, but to have the title: 'Blue Riband' was to be famous.

RMS Titanic

By the early 1900s travelling by ship was not unusual for the rich. Activities and games were organised and later gyms and swimming pools appeared. One of the largest and most luxurious ships of the day belonged to Cunard's rival company, the White Star Line; it was called the *Titanic*. People were amazed at the size and beauty of the ship. They were told the *Titanic* couldn't sink, but in 1912 on her [4] first journey to America, the ship sank after hitting an iceberg. [5] 1517 people lost their lives. Investigators have asked questions about why

3. **prize**: a reward for winning.
4. **her**: the pronoun 'she' is used instead of 'it' for a ship.
5. **iceberg**: large area of frozen water in the sea.

the *Titanic* sunk. How did it hit the iceberg? Was it going too fast? We will never know all the answers for certain. Films have been made about the *Titanic* and this gives an interesting view of how the different classes travelled on the ship but not everything in the films is based on fact. Today the remains of the *Titanic* lie at the bottom of the sea; occasionally they are visited by professional divers.

Modern Transatlantic Travel

The 19th century saw the arrival of the aeroplane. Aeroplanes started carrying more passengers. They were faster and they became cheaper too. By the 1960s they were the main method of travel across the Atlantic. Depending on the cruise, a cruise ship today can take 1 week from the UK to America, and about 100 days to go around the world. People take cruises to enjoy the experience of travelling, which is more important than arriving at their destination quickly. For some people, travelling by ship reminds them of how it was to travel more than 100 years ago.

1 Comprehension check
Put these events in the order they happen in history.

1 ☐ The *Titanic* sinks after hitting an iceberg.
2 ☐ The *Britannia* takes passengers across the Atlantic.
3 ☐ Aeroplanes are the fastest way to travel to America.
4 ☐ The industrial revolution means bigger and better engines.

2 Poster
Look at the poster on page 88. Write an advertisement for either a modern cruise ship, or a ship at the time *Around the World in Eighty Days* was written, telling people:

• where the ship travels to;
• what facilities there are on the ship.

Before you read

1 Listening

Listen to the first part of Chapter Eight. Complete the text with the words you hear.

None of the (**1**) .... in the (**2**) .... of New York were leaving before 14 December: too (**3**) .... to arrive at the Reform Club before eight forty-five on 21 (**4**) ....

Passepartout was very upset. They were only forty-five minutes late and he was certain it was his (**5**) .... Phileas Fogg did not want Passepartout to feel responsible and he simply said, 'We'll see what happens (**6**) ....'

The next day at midday on 12 December, with just (**7**) .... days thirteen hours and forty-five minutes to return to London, Phileas Fogg went to look for a ship — large or small — to take them (**8**) .... the Atlantic.

2 Vocabulary

Which of these words can/can't you see in the photos below? In 5 minutes write as many words as you can connected to 'Ships and the Sea'.

cabin   coal   pirate   cargo   wheel   crew

CHAPTER **EIGHT**

The captain of a ship

11 December

None of the boats in the port of New York were leaving before 14 December: too late to arrive at the Reform Club before eight forty-five on 21 December.

track 15

Passepartout was very upset. They were only forty-five minutes late and he was certain it was his fault. Phileas Fogg did not want Passepartout to feel responsible and he simply said, 'We'll see what happens tomorrow.'

The next day at midday on 12 December, with just nine days, thirteen hours and forty-five minutes to return to London, Phileas Fogg went to look for a ship — large or small — to take them across the Atlantic.

end

After trying several ships with no success, he spoke to the captain of a cargo ship, the *Henrietta*.

'When is the ship leaving?' he asked the captain.

'In an hour,' he replied.

'Where is the ship going to?' Phileas Fogg asked.

'To France. Bordeaux.'

'Will you take myself and three other passengers to Liverpool?'

'To Liverpool? Certainly not,' said the captain, looking at him like he was mad. 'This ship must arrive in Bordeaux by 20 December.'

Phileas Fogg thought for a moment. 'I'll give you two thousand dollars for each passenger if you take us to Bordeaux then,' he said.

'Two thousand dollars each?' he repeated, amazed at such a generous offer. He scratched [1] his head. Why was this man offering him so much money? Did he have something to hide? It was a lot of money.

He agreed. 'The ship leaves at nine,' he said.

Two hours later our four travellers were on the *Henrietta* and they were leaving the port of New York on their way to Bordeaux.

The next day, on 13 December, Phileas Fogg was the new captain of the ship, and the *Henrietta* was going to Bordeaux.

However, Phileas Fogg gave the crew some money and they agreed to go to Liverpool instead. The crew locked the captain in his cabin, and he was now in there shouting and trying to free himself unsuccessfully. His companions were surprised to find that Phileas Fogg was a good sailor. Passepartout tried to ask him about this, but Phileas Fogg did not want to talk. He had to

---

1. **scratched** : touched it with his fingernails.

try and cross the Atlantic in stormy weather and his mind was concentrating on arriving in England on time. He also did not want to lose the ship and all its crew!

Fix decided that Phileas Fogg was not just a thief, he was a pirate. 'He's not taking the ship to Liverpool,' he thought, 'but to some unknown place where it's safe for him to escape. If I don't get help from the police there, all is lost!'

They were half-way across the Atlantic with only five days to go. They were going at top speed and everything was going well, until one of the men came to speak to Phileas Fogg.

'If we continue at this speed, we won't have enough coal to get the ship to Liverpool,' he said. 'We must slow down!'

'We can't,' replied Phileas Fogg. 'We'll burn all the wood on the ship, if we have to.'

When they came close to Ireland, only the outside metal of the body of the ship remained. The ship could not get to Liverpool, so they stopped in Ireland.

Phileas Fogg took the ship to a port called Queenstown. From there they took a train to Dublin and then a ship to Liverpool.

Inspector Fix did not understand the man. What was in Liverpool? What was in London that he wanted to return for? This voyage seemed endless.

At twenty to midnight on 20 December, they finally arrived in Liverpool. They were exactly six hours away from London. Enough time to get to the Reform Club to win the bet. Inspector Fix put his hand on Phileas Fogg's shoulder.

'You are Mr Phileas Fogg. Is that correct?' he said.

'Yes,' said Phileas Fogg slowly. He was a little confused by the question.

'Phileas Fogg, I am arresting you in the name of the law,' [2] he said. In a moment two policemen stood next to Phileas Fogg.

Phileas Fogg was very angry. 'You...!' he began. 'I don't like men who have no loyalty to people who help them like his own friends. How could you do this, when I thought you were an honest person? You are worse than a criminal!'

For the first time Inspector Fix felt bad. He did not know what to do now that he did not have to follow Phileas Fogg around the world. He could see that he was not really a bad man. But... he was a thief, and he wanted the reward for his efforts. He went out of the room. A policeman took Phileas Fogg away and Mrs Aouda started crying loudly. She put her head on Passepartout's shoulder and they left.

Phileas Fogg looked at the walls. He had no money. All his hopes were gone! He was in a police station because they thought he was a thief. He was losing a lot of money. He could only hope for one more amazing event to rescue him.

It came sooner than he thought.

Passepartout ran back to the police station with Mrs Aouda. Then Inspector Fix arrived too. His hair was untidy and he could not breathe.

'Mr Fogg,' he cried when he could finally speak. 'Mr Fogg! You are free to go! They caught the thief three days ago!'

2. **I am arresting you in the name of the law** : policemen say this before they take someone to the police station.

The text and **beyond**

1 Comprehension check

There are some incorrect words in the sentences (1-8). Find the incorrect words and replace them with a word (A-H) from the box. Then answer the questions. (You can use exercise two to help you.)

> A fault   B free   C travel   D arrived   E ships
> F captain   G passenger   H crew

1 ☐ Could any of the trains take them to England on time?
2 ☐ Why did Passepartout feel certain it was his money?
3 ☐ Where did the police need to be on 20 December?
4 ☐ How much did Fogg pay for each bag?
5 ☐ Where did the pirates lock the Captain of the ship?
6 ☐ Why can't they escape as far as Liverpool?
7 ☐ What happened as soon as they stayed in Liverpool?
8 ☐ Why did the police say that Fogg was happy to leave?

PET 2 Sentence transformation

For each sentence complete the second sentence so that it means the same as the first. Use no more than three words. The first is done for you as an example.

0 They did not have enough time to get there.
They were not going to arrive on time.

1 He was sure that it was his fault.
He .... that he was to blame.

2 By 20 December the ship must arrive in Bordeaux.
The ship has .... in Bordeaux by 20 December.

3 "You'll receive $2,000 dollars for each passenger."
"I .... you $2,000 dollars for each passenger."

4 The crew locked the captain in his cabin.
The captain .... in his cabin by the crew.

97

5   If we continue at this speed, we won't have enough coal.
    There .... enough coal, if we continue to travel so fast.

6   The police took Phileas Fogg to the police station.
    The police told Phileas Fogg .... the police station with them.

3 **Characters**

What does Inspector Fix do in the story that can be described with these adjectives? Write your ideas below.

confused   greedy   determined   clever

....

....

....

4 **Captain, Pirate or Gentleman?**

Look at the notes Fix makes about Fogg on the ship. With another student choose an identity (captain, pirate, etc.) for Fogg and say why you think this is his true identity.

Thief behaves like a gentleman. Gentleman likes sailing but how did he take control of the crew/ship? Fogg was a Captain? Not v. interested in places he visits but knows about other countries. Mysterious. Is he just a thief? Why is he so rich? Can use a gun. A Pirate!

Example: Fogg is a pirate because he can sail a ship, he can
           fight ..., etc.
           Fogg is a cowboy because ...

5 **Writing**

Imagine you are a police officer. Record the details of Fogg's arrest, including: suspect, place, date, time, crime, what happens.

arrested ..... in the city of .... on ... at ..... on suspicion of ....

Before you read

1 **Reading pictures**
Look at the picture on page 103 and answer these questions:

1 What do you think Mr Fogg is telling Mrs Aouda?
2 How would you describe the expression on her face?
3 Why is Passepartout looking so happy?

Look at the picture on page 105 and answer the following questions:

1 Who are the people in the picture?
2 Where are they?
3 What time is on the clock? What does this mean?

2 **Predictions**
Answer the following questions in small groups.

1 What do you think the title means?
2 Do you think Phileas Fogg will win his bet? In pairs, choose different opinions and say why Fogg is/isn't successful.
3 Read exercise 3. Try to predict which sentences are true before you listen.

3 **Listening**
Listen to the first part of Chapter Nine and say whether the sentences below are true (T) or false (F).

	T	F
1 The Reform Club members believe Fogg has been successful.	☐	☐
2 Inspector Fix stops Phileas Fogg from leaving.	☐	☐
3 They don't arrive in London on time.	☐	☐
4 At 11.30 Phileas Fogg doesn't go to the Reform Club.	☐	☐
5 Mrs Aouda says she wants to return to India.	☐	☐
6 Phileas Fogg says he has no relatives and no friends.	☐	☐

track 16

When it is better to travel east

At the Reform Club, they held their breath. It was now almost eighty days since Phileas Fogg's departure and there was no news about his return. They read in the newspapers of an arrest for the bank robbery in Edinburgh — a respectable gentleman. But what about Fogg? Was he still on his journey? Was he still alive? They did not think he was going to walk through the door, but then with Phileas Fogg, you never knew.

track 16

Now Phileas Fogg was free and he knew exactly what to do. He looked at Inspector Fix and then he hit him: first with one hand, then with the other. Fix fell to the floor. Passepartout was very happy. 'Good!' he told his master. Then he turned to Inspector Fix.

'That's what happens to people who behave like you,' he shouted at the confused detective. They left the police station and went towards the railway station immediately.

Phileas Fogg knew the difficulties ahead, one minute here or there could change everything. They were in time for the train, but the train was late and when they arrived in London, they looked up at the clock in Euston Station to see that it was ten to nine. All was lost. They were five minutes too late to go to the Reform Club.

Phileas Fogg lost his bet and there was nothing he could do.

He accepted this in his usual way, without showing any particular emotion. Mrs Aouda, on the other hand, was very emotional. She continued to cry. She did not know what to do. Passepartout was also worried for his master, and his job. It was his master's choice to spend all his money on the bet, but he was such a good, honest person. It was not good to see him like this. He still could not stop thinking that it was his fault.

The next day Passepartout followed the same routine, except for one thing. When they heard the sound of Big Ben [1] at half past eleven the next day, Phileas Fogg did not go to the Reform Club.

The house felt strange. It was like no one lived there.

At about half past seven that evening Phileas Fogg asked Mrs Aouda if he could come to her room to speak to her.

'Madam,' he began sadly. 'I wanted to take you back to England with me because I thought I could offer you a good life here. Now I am a poor man... and I have nothing to offer you.'

---

1. **Big Ben** : clock in the tower of the Houses of Parliament in London.

It was the first time Mrs Aouda saw Phileas Fogg looking really sad.

'I know, Mr Fogg, and I'm so sorry. You saved my life. You took time to rescue me, and you lost your bet because of me.'

'Madam, I couldn't let you die that terrible death, but now you are here, and you need a way to live. I have my house and my possessions...'

'But what about you?'

'I don't need anything.'

'Maybe your friends could...'

'I have no friends,' he said sadly.

'Well, what about your relatives?'

'I have no relatives.'

end

'It is easier to live in poverty when there are two of us to share [2] it,' said Mrs Aouda taking his arm. 'I want to be your wife.'

Mr Fogg got up. Mrs Aouda saw that there was a small tear in his eye.

'I love you,' he said. 'And I want to spend my life with you.'

'Oh..!' said Mrs Aouda with a surprised cry. She was so happy!

Passepartout came into the room and saw his master standing close to Mrs Aouda. He understood immediately.

'This is wonderful news!' he said, 'We all need some good news.'

'Yes,' said Phileas Fogg, 'If you agree Mrs Aouda, we can get married immediately. Passepartout, do you know where Reverend Wilson lives?'

Passepartout ran to Reverend Wilson's house, but five minutes later, at twenty-five to eight he was already back at the house.

2. **share** : (here) to live the same life together.

'Tomorrow morning...' he said out of breath. 'You can't get married!'

'Why?' asked Phileas Fogg.

'Because today is Saturday and tomorrow is Sunday!' he said excitedly.

'Saturday? Impossible!' replied Phileas Fogg.

'Yes, yes it is. Do you remember? We went around the world and we travelled east and time changes as you go around the world and we're now twenty-four hours ahead. It's Saturday! Hurry, Mr Fogg! We only have ten minutes. You can still win your bet.'

They took Phileas Fogg's carriage to go to the Reform Club. Passepartout wanted to drive. He almost hit two dogs and they almost had more accidents before they arrived at the Reform Club at eight forty-four. Phileas Fogg's friends were waiting around the table, counting the seconds.

'Well, hello my friends,' he said, 'I believe that I am now a rich man,' he said with a small smile when he stepped into the Games Room at eight forty-five.

They all agreed. Here he was, eighty days later.

And that was how Phileas Fogg won his bet.

On Monday morning Phileas Fogg and Mrs Aouda were married. Later that morning Passepartout came into his room.

'Do you know, Mr Fogg,' he said, 'I read that if you don't go across India, you can go around the world in just seventy-eight days?'

'Maybe that's true,' said Phileas Fogg. 'But when we went across India, I met Mrs Aouda, who is now my lovely wife.'

And with these words they celebrated Phileas Fogg's good fortune.

The text and **beyond**

1 **Comprehension check**

Put the sentences about Chapter Nine in the correct order.

1 ☐ Phileas Fogg wins his bet at exactly 8.45.
2 ☐ Passepartout goes to Reverend Wilson's house.
3 ☐ Mrs Aouda agrees to marry Phileas Fogg.
4 ☐ Fogg loses his bet by just five minutes.
5 ☐ Mrs Aouda and Phileas Fogg become husband and wife.
6 ☐ Fogg is so angry he hits Inspector Fix.
7 ☐ Everyone is sad about the lost bet.
8 ☐ They realise they are one day ahead.

2 **Spot the difference**

Spot the odd one out and write your reasons in the lines below.

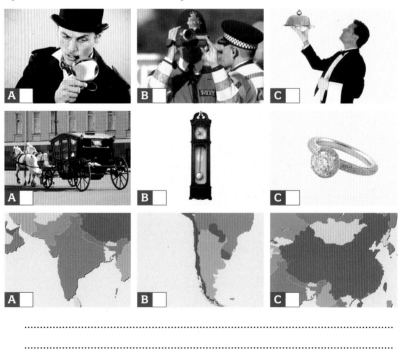

....

....

....

3 Time zones

Find the time differences between these countries:

Mumbai and Singapore; London and New York; Hong Kong and Sydney.

4 Game

Write the name of twenty countries on separate pieces of paper and fold them. Mix them together. In pairs open two each. The person with the greater time difference between countries wins. (No difference is a draw.)

5 Your trip around the world

You have come back home after your Around the World trip. Write four things you are going to do now that you are back in your own country.

Keep in touch!

List as many ways as you can use to keep in touch with English-speaking friends abroad.

PET **6** Writing

You receive an email from an English-speaking friend you met when you were travelling. Write a reply. (approximately 100 words)

Hi!

Are you back home now? What's the weather like? What did you do when you arrived? I remember you missed the food when you were travelling. What are you going to do now you're back? Are you going to come and visit again?

I hope one day I can come and visit you. Keep in touch!

Madi X

7 Keep a diary

Task 1: Keep an English diary. You can keep a record of new things that you learn or simply practice writing about daily/weekly events in English.

Task 2: Make a list of information you would like to update for other people to read. Find out how to create a blog. (If you already use a social network site, add your English-speaking friends.)

Note: it is very important when using websites, blogs, forums, etc. to know how to protect your privacy and personal safety. If you are in a class, your teacher will explain this in more detail.

8 Speaking: the hot seat

A chair in front of the class is the 'hot seat'. While you are sitting in this chair you are Phileas Fogg or Mrs Aouda or Passepartout: you must answer as if you were one of these characters. Take turns sitting here. The rest of the class can ask the person in the hot seat any questions. Here are some examples. Then think of some other questions.

Questions for Phileas Fogg

Are you proud of yourself and your succesful journey around the world?

Did you ever think you wouldn't have made it?

How was travelling with Passepartout and Mrs Aouda?

Did you ever risk your lives?

Would you do it again?

What are you going to do next?

Questions for Passepartout

Did you ever find the journey stressful?

Which part of the journey was the most difficult and why?

How was travelling with Mr Fogg and Mrs Aouda?

Which part of the world would you like to go back again and why?

What was the most exciting adventure?

Are you still going to work for Mr Fogg?

Questions for Mrs Aouda

What was your reaction when you woke up between Passepartout and Phileas Fogg.

Did you love your husband?

Were you ever scared during the journey? If so, when?

What do you think of London?

Do you miss your country?

Are you happy to be Mr Fogg' s wife?

1 Picture summary

Look at the pictures from *Around the World in Eighty Days* below. They are not in the right order. Put them in the order they appear in the story and then write a line under each picture to summarise what happened in the chapter.

A

....

B

....

C

....

D

....

E

....

F

....

G

....

H

....

I

....

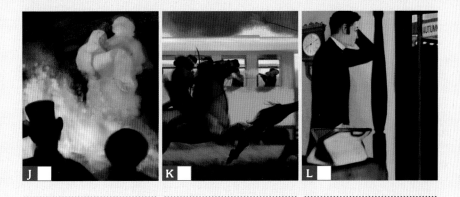

2 A graphic novel

Photocopy these two pages, cut out the pictures and stick them on paper in the right order. Think of words to put in speech or thought bubbles to show what the characters are saying or thinking. Do not use the words that were used in this book! Then write at least one sentence under each picture to narrate what is happening.

This reader uses the **EXPANSIVE READING** approach, where the text becomes a springboard to improve language skills and to explore historical background, cultural connections and other topics suggested by the text.

The new structures introduced in this step of our **READING & TRAINING** series are listed below. Naturally, structures from lower steps are included too. For a complete list of structures used over all the six steps, see *The Black Cat Guide to Graded Readers*, which is also downloadable at no cost from our website, blackcat-cideb.com.

The vocabulary used at each step is carefully checked against vocabulary lists used for internationally recognised examinations.

Step Two , B1.1

All the structures used in the previous levels, plus the following:

Verb tenses

Present Perfect Simple: indefinite past with *yet*, *already*, *still*; recent past with *just*; past action leading to present situation

Past Perfect Simple: in reported speech

Verb forms and patterns

Regular verbs and most irregular verbs

Passive forms with *going to* and *will*

So / neither / nor + auxiliaries in short answers

Question tags (in verb tenses used so far)

Verb + object + full infinitive (e.g. *I want you to help*)

Reported statements with *say* and *tell*

Modal verbs

Can't: logical necessity

Could: possibility

May: permission

Might (present and future reference): possibility; permission

Must: logical necessity

Don't have to / haven't got to: lack of obligation

Don't need to / needn't: lack of necessity

Types of clause

Time clauses introduced by *when*, *while*, *until*, *before*, *after*, *as soon as*

Clauses of purpose: *so that*; *(in order) to* (infinitive of purpose)